CELL BIOLOGY RESEARCH PROGRESS

PROSTAGLANDINS: BIOCHEMISTRY, FUNCTIONS, TYPES AND ROLES

CELL BIOLOGY RESEARCH PROGRESS

Additional books in this series can be found on Nova's website at:

https://www.novapublishers.com/catalog/index.php?cPath=23_29&seriesp=
Cell+Biology+Research+Progress

Additional e-books in this series can be found on Nova's website at:

https://www.novapublishers.com/catalog/index.php?cPath=23_29&seriespe=
Cell+Biology+Research+Progress

CELL BIOLOGY RESEARCH PROGRESS

PROSTAGLANDINS: BIOCHEMISTRY, FUNCTIONS, TYPES AND ROLES

GILLIAN M. GOODWIN
EDITOR

Nova Science Publishers, Inc.
New York

For permission to use material from this book please contact us:
Telephone 631-231-7269; Fax 631-231-8175
Web Site: http://www.novapublishers.com

Library of Congress Cataloging-in-Publication Data

Prostaglandins: biochemistry, functions, types, and roles / editors, Gillian M. Goodwin.
p. ; cm.
Includes bibliographical references and index.
ISBN 978-1-61668-272-9 (hardcover)
1. Prostaglandins. I. Goodwin, Gillian M.
[DNLM: 1. Prostaglandins. QU 90 P96125 2010]
QP801.P68P72375 2010
612.6'1--dc22

2010004376

Published by Nova Science Publishers, Inc. ✢ New York

CONTENTS

PREFACE

A prostaglandin is any member of a group of lipid compounds that are derived enzymatically from fatty acids and have important functions in the animal body. Every prostaglandin contains 20 carbon atoms, including a 5-carbon ring. They are mediators that participate in the regulation of multiple biological processes, both in health and disease. Prostaglandins contribute to the regulation of vascular tone, platelet function, and fertility. They also play key roles as inflammatory mediators and modulators of tumor biology. Although they are technically hormones, they are rarely classified as such. This book examines the biological role of prostaglandins in drug discovery, and on immune cells; the prostaglandin interchange; and the roles of prostaglandin E2 in inflammation-related cardiovascular disease as well as other related topics of import.

Chapter 1 - Cyclopentenone prostaglandins (cyPG) are reactive lipid mediators, derived from unsaturated fatty acids, which display varied biological actions. Interest in cyPG arose from early studies showing anti-viral and anti-tumoral effects of these compounds. The identification of certain cyPG as agonists of cellular receptors elicited a renewed interest in their study. More recently, an important role of cyPG in the modulation of the inflammatory response has been put forward. For these reasons, cyPG have become the object of intense study as potential therapeutic agents. Research into the mechanisms of the biological effects of cyPG has revealed that these compounds exert multi-faceted actions. An important property of cyPG is their ability to form covalent adducts with cysteine residues in cellular proteins, thus altering their activity. As a growing number of protein targets for modification by cyPG are being identified and the functional consequences of cyPG addition unveiled, it is becoming clear that this is a key mechanism for the potential beneficial effects of these prostanoids. The direct covalent modification and inhibition of several components of the NF-kB pathway contribute to cyPG anti-inflammatory effects, whereas addition to redox sensitive proteins plays a key role in the regulation of the cellular redox status and stress response. Therefore, cyPG are emerging as pleiotropic mediators, much like other endogenous molecules able to chemically modify proteins, like nitric oxide or reactive aldehydes, like 4-hydroxy-nonenal. However, in this complex scenario there is room for selectivity and regulation. Proteomic studies have revealed that cyPG target only a subset of cellular proteins. Moreover, the nature of the proteins modified by cyPG depends on the structure of the prostanoid and the protein, in such a way that cyPG with different structure may target distinct cellular proteins or even different cysteine residues within a given protein. This paves the way for structure-function studies and for the search of cyPG derivatives with increased selectivity or potency. In this chapter the authors will review the current advances in the cyPG field in the light of their recent findings.

Chapter 2 - Prostaglandins (PGs) are major regulators of growth and transport in kidney tubule epithelial cells. Although the kidney is a major producer of prostaglandins, little is understood about prostaglandin effects on tubule epithelial cells, and the signaling mechanisms involved in these effects. The authors approach has been the use of in vitro kidney tubule epithelial cell culture systems in hormonally defined, serum free medium. The authors have found that Prostaglandin E_1 (PGE_1) and PGE_2 stimulate both growth and ion transport by an established kidney epithelial cell line, Madin Darby Canine Kidney, a distal tubule model. Similar affects on growth were observed with primary baby mouse kidney epithelial cells, indicating their physiologic significance. Subsequent to these studies, the overproduction of PGE_2 in the tumor microenvironment has been found to contribute to the formation of a number of carcinomas. This is because PGE_2 stimulates the growth of many types of epithelial cells in vivo. The authors have also found that PGE_1 and PGE_2 stimulate the Na,K-ATPase in MDCK cells, as well as primary rabbit kidney proximal tubule cells. Regulatory effects of PGE_1 and PGE_2 on Na,K-ATPase were attributed to transcriptional control. Prostaglandin Response Elements (PGREs) were identified on the Na,K-ATPase beta subunit promoter. These PGREs are highly conserved elements present on genomic DNA in a number of species. In addition, the authors have identified a novel regulatory mechanism for gene expression via these PGREs. Both cyclic AMP Regulatory Element Binding Protein (CREB) and Sp1 bind to PGREs, and interact so as to modulate the rate of transcription. The molecular mechanisms underlying the CREB/Sp1 interaction are being defined.

Chapter 3 - Tissues of certain holothurians (Echinodermata, Holothuroidea, Dendrochirota) are known to contain relatively high levels of prostaglandins (PGs) including PGE_2. In contrast to vertebrates, the contributions of PGs to immunoregulation in echinoderms have been little explored. In this study, the authors investigated the specifics in effects of PGE_2 on apoptosis in different types of immune cells in holothurian *Eupentacta fraudatrix*. The studies were conducted in different cell fractions obtained after density gradient centrifugation: the fraction of phagocytes of higher density (P1), the fraction of phagocytes of lower density (P2), the fraction combined of two phagocyte types (CP), and fraction enriched in morula cells (MC). DNA fragmentation was detected by agarose gel electrophoresis and by Hoechst 33342 staining. All fractions studied demonstrated pronounced apoptotic background with specific intensity for each of them. Inhibitor of endogenous PG synthesis indomethacin (10 µg/g) administration one day prior to experiments inhibited apoptosis compared to its background value both in PC and MC. While indomethacin was used both in vivo (10 µg/g) and in vitro (10^{-6} M), the level of apoptosis insignificantly increased in freshly isolated MC, but not in PC. In 24 h incubation, PGE_2 (10^{-6} M) counteracted effects of indomethacin on apoptosis in MC but not in PC. In 48 h, both indomethacin (10^{-6} M) and the more combined indomethacin (10^{-6} M) plus PGE_2 (10^{-7}-10^{-6} M) decreased apoptosis both in MC and PC of animals pretreated with indomethacin in vivo. Dexamethasone (10^{-4}-10^{-7} M) induced apoptosis in the reverse concentration-dependent manner in PC in 24 h, and 10^{-4} M dexamethasone even inhibited apoptosis compared to the control. However, in P1, dexamethasone (10^{-4} M) significantly increased apoptosis compared to the control in 24 h. In 48 h, dexamethasone produced similar effects in P1 and MC, and the reverse dexamethasone-concentration dependence of apoptosis level was shown in both types of the cells. Commercially available catalase benefited apoptosis-modulating effects of both indomethacin and dexamethasone on apoptosis in fractions of phagocytes, but not in MC. Both indomethacin or dexamethasone (10^{-4} M) counteracted apoptosis-increasing effect of PG in normal MC, but failed to effect apoptosis reduction following PGE_2 (10^{-6} M) treatment in phagocytes. Apparently, PGE_2 may be involved in apoptosis modulation both in MC and phagocytes of holothurian *E. fraudatrix,* and different sensitivities of MC and phagocyte fractions to apoptosis-modulating effect of PG may play an important role in cooperation of immunocytes during immune response similar to that in vertebrates.

Chapter 4 - Previous studies using cells and tissue culture have demonstrated that prostamide and prostanoid FP receptor activities may be pharmacologically differentiated by selective prostamide receptor antagonists. The intent of the studies described herein was to reproduce this activity profile in the eyes of living animals. For this purpose, a second generation prostamide antagonist (AGN 211336) was used. AGN 211336 has a pA_2 of 7.6 at prostamide receptors but no meaningful activity (k_b>10,000 nM) at prostanoid FP and DP_1, DP_2, EP_{1-4}, and IP receptors. AGN 211336 also inhibited bimatoprost induced Cyr 61 mRNA upregulation in cells expressing the human recombinant prostamide receptor but did not block the response to $PGF_{2\alpha}$. The effects of AGN 211336 on bimatoprost and latanoprost induced ocular hypotension and ocular surface hyperemia were evaluated in beagle dogs over a five day experimental period. Pretreatment with AGN 211336 (2x5% suspension daily) statistically significantly inhibited the decrease in intraocular pressure produced by bimatoprost (0.03% daily). In contrast, an identical dosing regimen for AGN 211336 did not alter the ocular hypotensive response to latanoprost (0.03% daily). Thus, the prostamide bimatoprost and the prostanoid FP receptor agonist prodrug latanoprost behaved as pharmacologically discrete entities in the eye, according to their differential susceptibility to AGN 211336. The ocular hypotensive effects of bimatoprost were not blocked by the FAAH inhibitor OL-92 or a close structural congener of AGN 211336 that lacked prostamide antagonist properties (AGN 220675). Interestingly, AGN 211336 did not block the ocular surface redness produced by either latanoprost or bimatoprost. This suggests that the ocular surface hyperemia produced by bimatoprost occurs as a result of non-prostamide, off-target activity.

Chapter 5 - Phosphatidylserine (PS), which is normally located on the inner leaflet of the plasma membrane, translocates to the outer leaflet during early stages of apoptosis,. The externalization of PS provides a signal for phagocytes to initiate uptake of apoptotic cells. After the phagocytosis of apoptotic cells, phagocytes start to secrete anti-inflammatory mediators including prostaglandin E_2 (PGE_2), which act an autocrine manner to control their functions. PS-containing liposomes (PS liposomes) can mimic the effects of apoptotic cells on phagocytes to induce the secretion of PGE_2. PS liposomes induced PGE_2 secretion from microglia without induction of either cyclooxygenase (COX) 2 or microsomal prostaglandin E synthase (mPGES)-1. PS liposomes were found to utilize the COX-1/mPGES-2 system to produce PGE_2 secretion and then shift microglia and macrophages from a pro- to anti-inflammatory phenotype by the autocrine action of PGE_2. Moreover, PS liposomes inhibit the maturation of dendritic cells and osteoclast precursors. Therefore, PS liposomes may be potentially useful as one type of pharmacological intervention for inflammatory and immune diseases through a feedback mechanism involving PGE_2.

Chapter 6 - *Background:* The aggregation of platelets by aggregating agents is essential for normal blood coagulation process. In contrast excessive platelet aggregation is reported to result in the coronary artery disease (CAD). The homeostasis is achieved through the inhibition of platelet aggregation by several humoral factors including prostacyclin (PGI_2) and insulin. Prostacyclin, a potent inhibitor of platelet aggregation, is however without thrombolytic effect. Insulin, also a potent inhibitor of platelet aggregation is known to mediate its effect through synthesis of nitric oxide (NO) which is a potent inhibitor of platelet aggregation and a thrombolytic agent. The authors have reported before that the treatment of platelets with PGI_2 increased the insulin receptor density on the platelet surface without receptor synthesis; due to the exposure of the "spare" insulin receptors on the membrane from the membrane bilayer. The effect of the increased insulin receptor numbers in platelets due to the "cross talk" with PGI_2 receptors on the antithrombotic effect of insulin was determined.

Methods: The binding of the ligands [^{125}I-insulin and ^{3}H PGE_1, a stable probe for PGI_2 receptor] was determined by Scatchard plot analysis of the equilibrium binding of the agonists using membrane filtration technique. Nitric oxide was determined by methemoglobin method. Thrombolysis *in vitro* was determined by clot lysis spectrophotometrically.

Results: The treatment of platelets in Tyrod's buffer with 200μunits of insulin in the presence or absence of 2nM PGE_1 for 2 h (equilibrium binding for insulin) resulted in the increase of insulin receptor numbers (n) from ≈120 sites/platelets in the control to 250 sites/platelets that increased the formation of NO from 1.2 nmol/mg/h in the control to 2.5nmol/mg/h in the presence of PGE_1 which resulted in the increased rate of clot lysis by ≈2 fold compared to control.

Conclusion: The systemic increase of PGI_2 might increase the insulin binding to platelets through the increase of insulin receptor numbers on the cell surface. The increased binding of insulin would not only inhibit platelet aggregation, but also enhance the thrombolytic effect of the hormone *in vivo*.

Chapter 7 - Non-steroidal anti-inflammatory drugs (NSAIDs) are extensively used for the management of inflammation, pain, swelling and stiffness associated with diseases such as rheumatoid arthritis, osteoarthritis, inflammatory arthropathies, post-operative pains, etc. NSAIDs act via inhibition of cyclooxygenases (COX) resulting in subsequent inhibition of prostaglandins (PGs) in the arachidonic acid (AA) metabolic pathway. Traditional non-selective NSAIDs including indomethacin, ibuprofen and naproxen, though effective in the management of pain and inflammation, were associated with a number of gastrointestinal (GI) and renal side effects because of their ability to inhibit activities of both isoforms of cyclooxygenase, (COX)-1 and -2, thus resulting in inhibition of PGs and thromboxane (TXs) production. Since COX-1 regulates GI, renal and vascular functions, selective inhibition of inducible COX-2 was thought to be a better alternative and as a result selective COX-2 inhibitors (coxibs) were developed. Coxibs including celecoxib, rofecoxib and valdecoxib were claimed to produce anti-inflammatory effects with low GI side effects. Although coxibs were associated with reduced incidence of serious GI side effects, these drugs demonstrated an increased risk for cardiovascular events associated with the inhibition of COX-2. For this reason, rofecoxib in late 2004 and valdecoxib in 2005 were withdrawn from the market. This withdrawal has raised several serious questions regarding the balance of safety and efficacy of this class of drugs and, as a result, alternative targets in the AA metabolic pathway are being identified to overcome the side effects associated with coxibs. Since the discovery of microsomal prostaglandin E synthase-1 (a PGE_2 inducible enzyme downstream of COX), several reports have suggested the possibility of mPGES-1 as a future therapeutic target in counteracting inflammation during inflammatory diseases. This review will discuss briefly the prospect of mPGES-1 inhibition as an alternative therapeutic strategy for counteracting inflammation. The question that the authors will put in perspective is whether its pharmacological inhibition will bypass the cardiovascular side effects associated with COX-2 inhibitors?

Chapter 8 - The kinetic behavior of structures of prostaglandins A_1, A_2, B_1, B_2 and E_1, E_2 in ethanol is explored at 20-25°C by a method of a UV-spectroscopy. The absorbance readings were performed monthly within one year. It is shown, that prostaglandins are interchanged and the character of interchanges is individual. So, partial conversion of E_1 to A_1 needs one year. The conversion of E_2 to A_2 begins in three months. The intensive conversion of $E_2 \rightarrow B_2$ begins in one month, and in 12 months the equilibrium between E_2 and B_2 is established practically. At the same time reversible conversion of $B_2 \rightarrow E_2$ begins in 1 month, some more in 11 months, along with B_2, the authors have intensive peaks A_2 and E_2.

Chapter 9 - Prostaglandin E2 (PGE_2) is produced during inflammatory responses and controls a variety of immunological reactions through four different receptor subtypes (EP1-4) with various signaling cascades. However, the precise roles of each receptor in cardiovascular diseases have not yet been elucidated. It has been shown that PGE_2 suppresses myocardial ischemia and heart transplant rejection via its receptor EP4, suggesting a particular role of this subtype for myocardial inflammation. It has also been reported that enhanced expression of some subtypes of PGE_2 receptors has been shown in clinical and experimental cardiovascular diseases. However, their pathophysiological roles in the development of disorders remain unelucidated. Recently, the authors revealed the significance of EP4 for suppression of myocardial inflammation using a novel EP4 selective agonist (EP4RAG) in murine cardiac transplantation and myocardial ischemia models. The investigation indicated that EP4RAG attenuated inflammation by suppressing chemokines and the infiltration of macrophages. In this short review article, the authors are focusing on the roles of prostaglandin E2 receptors in the development of cardiovascular diseases.

Chapter 10 - Inflammatory responses are essential defense mechanisms which protect pluricellular organisms from various types of insults and imply the regulation of a vast number of genes. The transcription factors NF-κB and AP-1 are key players in inflammation, since they are rapidly activated by inflammatory stimuli and in turn direct the expression of inflammatory mediators such as cytokines, matrix metalloproteinases, iNOS and COX-2. The inflammatory response is a tightly controlled process. It possesses intrinsic mechanisms that limit its intensity and duration. Key inflammatory mediators such as nitric oxide and prostaglandins may contribute both to the onset and to the resolution of inflammation. Cyclopentenone prostaglandins (cyPG) are reactive compounds which may form covalent adducts with thiol groups in proteins or in glutathione. CyPG have proven potent antiinflammatory agents both in cellular and in animal models. CyPG may act as ligands for the nuclear receptors of the PPAR family of transcription factors, and PPAR-mediated responses appear to contribute to their beneficial effects. In addition, posttranslational protein modification by covalent cyPG binding may modulate signal transduction cascades leading to transcription factor regulation and/or affect the activity of transcription factors themselves. The transcription factors NF-□B and AP-1 have been identified as targets for this modification. In this work the authors have explored the effects of cyPG on the NF-κB activation pathway in mesangial cells, a known cellular model of inflammation. Activation of mesangial cells in the presence of the cyPG 15-deoxy-$\Delta^{12,14}$-prostaglandin J_2 reduces NF-□B transcriptional activity. Pre-treatment with cyPG inhibits NF-κB DNA binding activity both in intact cells and in isolated nuclear extracts. By using a biotinylated 15-deoxy-$\Delta^{12,14}$-prostaglandin J_2 derivative the authors have observed the modification of the NF-κB subunits p65 and p50 in intact cells. Subsequently, the authors have used biotinylated cyPG to study the interaction of these prostanoids with NF-κB in vitro. Binding of biotinylated cyPG to p50 is selectively reduced in the presence of an oligonucleotide containing the consensus site for NF-κB binding, which suggests that this interaction involves the DNA binding domain. Labeling studies of a cys62ser p50 mutant confirm that this cysteine residue is a site for cyPG attachment. Biotinylated cyPG labeling assays indicate that other proinflammatory transcription factors are potential targets for modification by cyPG. In addition, cyPG may modulate the activity of proteins along the activation pathways of these factors. Moreover, cyPG may alter cellular redox state by various mechanisms. Given the fact that numerous transcription factors, including NF-κB and AP-1 are redox-regulated, both direct and indirect mechanisms can be envisaged for the modulation of pro-inflammatory transcription factors by cyPG. Finally, a number of endogenous lipids as well as natural products possess reactive unsaturated carbonyl groups, the approaches used in this study could prove useful to assess their potential in the regulation of transcription factor activity.

Chapter 11 - There have been frequent suggestions in recent years that therapeutic benefits, in particular the maintenance of an adequate blood supply to the retina or optic nerve head (ONH), may results from the topical application of drugs.

Prostanoid analogues have recently been introduced into clinical use for the management of increased intraocular pressure (IOP).[1.2] This class of compound is known to exert effect on vascular components.[3.4] Some endogenous prostaglandins have been shown to influence regional ocular blood flow and exhibit significant vasoactive proprieties in ophthalmic, ciliary and retinal arteries and veins. [5.6.7.8] In glaucomatous eyes, therapies that negatively impact ocular blood flow could interfere with the effective transport of oxygen and nutrients to already compromised areas of retina.

The purpose of this review is to examine in detail the "state of art" of the prostaglandin-induced interaction with ocular blood flow in experimental models and in glaucomatous patients.

In: Prostaglandins: Biochemistry, Functions… ISBN: 978-1-61668-272-9
Editors: Gillian M. Goodwin, pp.1-25

Chapter 1

CYCLOPENTENONE PROSTAGLANDINS: SIGNALLING MECHANISMS AND OPPORTUNITIES FOR DRUG DISCOVERY

Dolores Pérez-Sala
Department of Chemical and Physical Biology, Centro de Investigaciones Biológicas, CSIC. 28040 Madrid, Spain

ABSTRACT

Cyclopentenone prostaglandins (cyPG) are reactive lipid mediators, derived from unsaturated fatty acids, which display varied biological actions. Interest in cyPG arose from early studies showing anti-viral and anti-tumoral effects of these compounds. The identification of certain cyPG as agonists of cellular receptors elicited a renewed interest in their study. More recently, an important role of cyPG in the modulation of the inflammatory response has been put forward. For these reasons, cyPG have become the object of intense study as potential therapeutic agents. Research into the mechanisms of the biological effects of cyPG has revealed that these compounds exert multi-faceted actions. An important property of cyPG is their ability to form covalent adducts with cysteine residues in cellular proteins, thus altering their activity. As a growing number of protein targets for modification by cyPG are being identified and the functional consequences of cyPG addition unveiled, it is becoming clear that this is a key mechanism for the potential beneficial effects of these prostanoids. The direct covalent modification and inhibition of several components of the NF-kB pathway contribute to cyPG anti-inflammatory effects, whereas addition to redox sensitive proteins plays a key role in the regulation of the cellular redox status and stress response. Therefore, cyPG are emerging as pleiotropic mediators, much like other endogenous molecules able to chemically modify proteins, like nitric oxide or reactive aldehydes, like 4-hydroxy-nonenal. However, in this complex scenario there is room for selectivity and regulation. Proteomic studies have revealed that cyPG target only a subset of cellular proteins. Moreover, the nature of the proteins modified by cyPG depends on the structure of the prostanoid and the protein, in such a way that cyPG with different structure may target distinct cellular proteins or even different cysteine residues within a given protein. This paves the way for structure-function studies and for the search of cyPG derivatives with increased selectivity or potency. In this chapter we will review the current advances in the cyPG field in the light of our recent findings.

1. Introduction

Prostaglandins (PG) are lipidic mediators that participate in the regulation of multiple biological processes, both in health and disease. PG contribute to the regulation of vascular tone, platelet function, and fertility. PG play also key roles as inflammatory mediators and as modulators of tumor biology. PG are synthesized from 20 carbon fatty acids, such as arachidonic acid by the action of enzymes known as cyclooxygenases (COX). COX are bifunctional enzymes that generate PGH_2. From this common precursor, a variety of synthases generate a plethora of PG of diverse structure. The multiple biological effects of PG arise from their structural diversity, from the tissue specific distribution of the various PG synthases and from the distribution and diversity of PG receptors. One of the examples of this diversity is provided by PGE_2, a pro-inflammatory PG that can induce contraction or relaxation of smooth muscle depending on the subtype of PGE receptor upon which it acts, with EP2 and EP4 being relaxant receptors and EP1 being contractile [1], and the tissue distribution of the different subtypes [2, 3]. A schematic view of PG synthesis is shown in Figure 1. In turn, PG can suffer additional transformations to generate PG derivatives with different biological activity. PG with cyclopentenone structure (cyPG) are generated by PG dehydration. These endogenous electrophiles have received considerable attention due to their varied biological effects and their therapeutic potential. However, whereas the pharmacological effects of these compounds are well established, their pathophysiological significance is not completely understood. In order to elucidate whether cyPG fulfil a pathophysiological role as lipid mediators, several features of these compounds need to be assessed. In the first place, they should be generated in vivo at concentrations sufficient to elicit biological effects. It would be also interesting to explore whether the effects of cyPG display selectivity and whether their actions are subjected to regulatory mechanisms, which could include reversibility of the modification of protein targets or feedback mechanisms. Several recent works shed light on these aspects of cyPG biology.

2. Formation of cyPG

cyPG are generated from other PG by non-enzymatic dehydration. PG of the A series arise from the dehydration of PGE-type PG, whereas cyPG o the J series are generated by dehydration of PGD_2. Among PGA-type cyPG, PGA_2 is derived from PGE_2 and PGA_1 arises from PGE_1, which is synthesized by the action of COX on dihomo-γ-linolenic acid [4]. Besides, it should be noted that both, PGE_2 and PGD_2, can also be formed in substantial amounts through the oxidation of arachidonic acid by non-enzymatic mechanisms via the isoprostane pathway [5].

CyPG of the A series were detected in various biological fluids more than thirty years ago [6, 7]. Early experiments assessing the biological effects of these compounds showed that they displayed potent anti-viral and anti-proliferative actions [8, 9]. During the 1980s several cyPG of the J series were identified in biological fluids or found to be generated in vitro from the dehydration of PGD_2. These include PGJ_2, identified in studies pursuing the mechanism for the antitumoral activity of PGD_2 [10], 15-deoxy-$\Delta^{12,14}$-PGJ_2 (15d-PGJ_2), which was found to arise from PGD_2 in the presence of albumin [11], and Δ^{12}-PGJ_2, detected after in vitro incubation of PGD_2 with human plasma [12], as well as naturally present in human urine [13]. Later, it has been found that PGD_2 suffers a dehydration reaction to yield PGJ_2, and this is further dehydrated to 15d-PGJ_2 in a process independent from the presence of albumin, whereas in the presence of albumin PGJ_2 it is converted to Δ^{12}-PGJ_2 [14] (Figure 1).

Structurally, cyPG are characterized by the presence of an α,β unsaturated carbonyl group in the cyclopentane ring (cyclopentenone) (Figure 2). This structure confers a strong electrophilicity to the carbon(s) in the -β position, which may suffer the nucleophilic attack from nucleophiles, mainly the sulphur atom in thiol groups, giving rise to Michael adducts. This leads to the formation of conjugates of cyPG with glutathione (GSH) or with proteins.

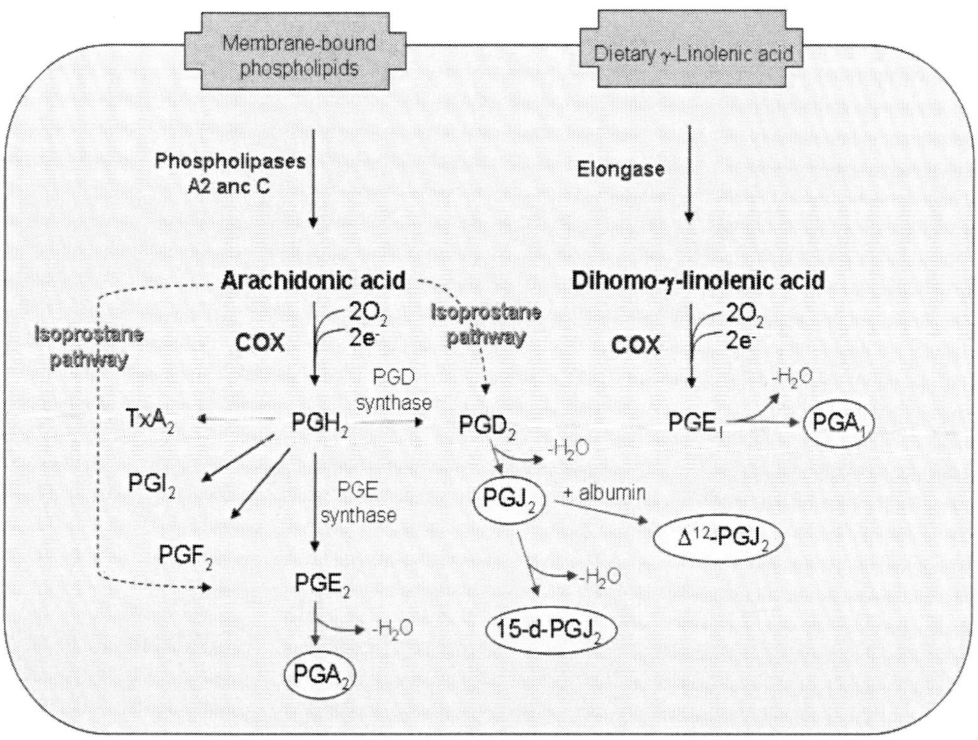

Figure 1. Pathways for the synthesis of cyPG. Enzymatic (COX-mediated) or non-enzymatic (isoprostane pathway) oxidation of polyunsaturated fatty acids can give rise to PG. Non-enzymatic dehydration of some PG leads to the formation of reactive cyPG (in yellow ellipses). Scheme has been adapted from several sources [1, 4, 14].

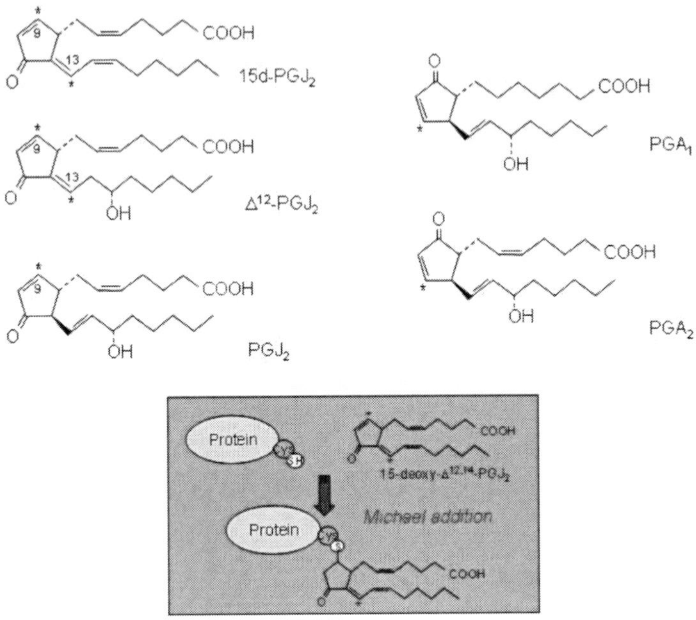

Figure 2. Structures of several cyPG and scheme of the Michael addition. Electrophilic carbons are marked by asterisks

3. Biological Effects of cyPG

The discovery of the anti-proliferative and anti-inflammatory effects of cyPG in various experimental systems, including in vivo models, has led to a burst in the number of works addressing the biological effects of these compounds. For this reason it is not possible to address here all the effects that have been described in various specific settings. Therefore, we will outline the better established actions. Early works in this field uncovered the anti-viral effects of these compounds. These effects were found to correlate with the ability of cyPG to induce a heat shock response and to inhibit the activity of the transcription factor NF-κB, which is necessary for the replication of numerous viruses [15]. In turn, the heat shock response leads to the expression of many cytoprotective proteins, including molecular chaperones like Hsp70, which play a protective role against several kinds of stress. Recently, it has also been shown that cyPG may target viral proteins important for transcriptional regulation [16].

The anti-proliferative effects of cyPG were evidenced both in vitro and in vivo, using cancer cell lines and in vivo models of carcinogenesis, respectively [9, 17]. In these works it was found that the cyclopentenone moiety was required for the growth inhibitory action of these compounds [18]. In the view of these results, a potential anti-tumoral activity of cyPG was proposed. Indeed, several formulations of cyPG have been assayed for their ability to inhibit tumor growth in vivo [19]. The anti-proliferative actions may arise from a variety of effects. CyPG have been found to alter cell cycle regulatory proteins [20-22]. Moreover, cyPG may induce profound alterations of cytoskeletal organization, which may affect vimentin, actin and/or tubulin distribution [23-26]. In addition, cyPG may induce potent oxidative stress by acting mainly on mitochondria [27, 28], although an action on other ROS generating systems, like NADPH oxidase, has also been reported [29]. cyPG induce apoptosis in a variety of cell lines through various mechanisms which, in addition to the inhibition of cell-cycle regulating proteins and/or the generation of ROS, may also involve the inhibition of pro-survival factors or of protein synthesis, as well as the interaction with pathways regulated by cell death receptors (see [30], for review).

cyPG are arising as key regulators of inflammation. In general, an anti-inflammatory role has been described for these compounds, although in some settings, cyPG may exert pro-inflammatory or dual actions. cyPG interact with pro-inflammatory pathways at multiple levels. 15d-PGJ_2 inhibits the induction of numerous genes, including iNOS, COX-2 or ICAM-1, elicited by pro-inflammatory stimuli [31, 32], reduces the release of inflammatory cytokines [33], inhibits monocyte trafficking through regulation of cell adhesion molecules and diminishes chemokine expression (reviewed in [34]). 15d-PGJ_2 has also been proposed to contribute to the resolution of inflammation through various mechanisms, including the induction of apoptosis in macrophages [35] and the regulation of the clearance of leukocytes from the site on inflammation [36]. Some of the potent anti-inflammatory effects of cyPG observed in several experimental systems stem from their ability to inhibit key pro-inflammatory transcription factors, including AP-1 and NF-κB [37-39], and to activate the transcription factors of the PPAR family. These, in turn, regulate macrophage function and may interfere with NF-κB and AP-1 signalling by various mechanisms, including competition for limiting amounts of coactivators, modulation of the expression of negative regulators (i.e.: by inducing the expression of IκB, the inhibitory subunit of NF-κB), direct protein-protein interactions and interference with DNA binding [40-43].

An important aspect of cyPG biology is their ability to regulate cellular redox status and the electrophilic response. As stated above, cyPG can induce potent oxidative stress, which may result in the oxidation of proteins. However, cyPG may exert a dual effect on cell redox status since 15d-PGJ_2 has been reported both to stimulate and inhibit GSH synthesis in a concentration-dependent fashion [44]. Moreover, cyPG activate the transcription factor Nrf2, which translates signals from electrophilic compounds and induces the expression of genes involved in the antioxidant response. In fact, a pre-conditioning effect of cyPG has been proposed by which pre-exposure to non-toxic concentrations of these compounds would increase cellular resistance towards further insults from oxidants or electrophilic agents.

4. Mechanisms of Action of cyPG

The diversity of the actions of cyPG arise from their pleiotropic mechanism of action. Until now, three main mechanisms of cyPG action have been identified, which may operate simultaneously in a given cell type.

4.1. Interaction with Nuclear Receptors

In the 90's the cyPG 15d-PGJ_2 was found to act as an agonist of the nuclear receptor PPARγ, thus being the first endogenous ligand identified for an, until then, orphan receptor [45, 46]. PPARγ is one of the members of the PPAR family of transcription factors, integrated also by PPARα and β/δ. All three PPARs function as heterodimers with the receptor of 9-cis-retinoic acid (RXR). Ligand binding elicits conformational changes that facilitate the interaction with transcriptional co-regulators. PPARγ was found to play an important role in lipid metabolism and inflammation. This generated great excitement and spurred the work on the effects of 15d-PGJ_2. Since then, a number of works have elucidated the structural basis for the interaction of 15d-PGJ_2 and other endogenous fatty acids with PPAR [47, 48]. These studies have shown that several endogenous PPARγ ligands, including 15d-PGJ_2, establish a covalent interaction with a cysteine residue located in the PPARγ ligand binding domain, C285, and elicit conformational changes is other parts of the molecule that are required for the appearance of the active conformation. These two processes are not always related since many PPARγ agonists do not bind covalently to the receptor and, conversely, some compounds able to bind covalently to C285 do not cause the conformational changes required for activation [49]. However, activation of PPARγ is not the only mechanism of cyPG action. This has been the reason why the use of cyPG and of 15d-PGJ_2 in particular, as tools to assess the involvement of PPARγ in various processes, has sometimes generated misinterpretations. In order to confirm the involvement of PPARγ in the effects observed, the action of cyPG has been compared in several studies with that of other PPAR agonists or antagonists. The fact that PPARγ knockout animals die before birth, has raised the need for conditional and tissue-specific knockout models to explore the role of these transcription factors (please see [50] for review). Results from these studies have confirmed that PPARγ plays a key role in adipocyte differentiation, inhibits macrophage pro-inflammatory responses and modulates macrophage function in multiple ways that can have an impact on glucose tolerance, skeletal muscle insulin sensitivity and anti-atherosclerotic effects [51]. However, the evidences arising from different experimental knockout models have sometimes yielded contradictory results, and more work is needed to ascertain the role of PPARγ in skeletal muscle metabolism and regulation of vascular tone. Another source of potential confusion in the PPARγ field is the fact that some PPARγ agonists may activate other receptors of the family. In fact 15d-PGJ_2 may exert biological effects such as nuclear translocation of all three PPAR [52] and several cyPG, including 15d-PGJ_2 and PGA_1 can activate PPARα and δ in several in vitro and cell-based assays [53, 54]. Finally, even classical PPARγ activators, like the thiazolidinediones may display PPARγ-independent effects. In summary, although it is clear that 15d-PGJ_2 is a potent agonist of PPARγ, the possibility of PPARγ independent effects needs to be taken into account.

4.2. Interaction with Membrane Receptors

The cyPG 15d-PGJ_2 and Δ^{12}-PGJ_2 have been reported to be ligands for membrane receptor DP2 [55, 56], one of the receptors of PGD_2, also known as CRTH2, because it was termed in the first place "chemoattractant-receptor homologous molecule expressed on Th2 cells". DP2 is a G-protein-coupled receptor, expressed in various types of leukocytes including eosinophils, basophils, some Th2 lymphocytes and monocytes, and it has been involved in chemotaxis and allergic inflammation (see [57] for review). However, the role of this receptor in inflammation is not completely elucidated. DP2 appears to play a positive role on leukocyte stimulation in vitro; however, knockout animals seem to be more susceptible to lung inflammation in certain experimental settings (reviewed in [57]). The interaction of 15d-PGJ_2 and DP2 occurs at nanomolar concentrations of 15-PGJ_2 and may result in calcium mobilization and chemotaxis of leukocytes [58] and migration of cells ectopically expressing the receptor [59]. However, the involvement of DP2 in the anti-inflammatory or pro-resolution effects of 15d-PGJ_2 is not clear. In a model of murine peritonitis 15d-PGJ_2 reduced peritoneal macrophage accumulation, increased numbers of macrophages in the parathymic lymph nodes and contributed to the balance of cytokines; however, specific DP2 agonists did not mimic these effects [36]. Activation of DP2 appears also not to be responsible for the anti-inflammatory and cystoskeletal disrupting effects of 15d-PGJ_2 in mesangial cells, since DP2 agonists such as indomethacin, do not mimic the effects of 15d-PGJ_2, which occur with micromolar concentrations of the cyPG [23]. Similarly, PGD_2 and 15d-PGJ_2 alter LPS and allergen-induced dendritic cell maturation by mechanisms independent from DP2 [60]. Therefore, studies with DP2 knockout animals will probably be needed to ascertain the role of this receptor in the effects of 15d-PGJ_2.

4.3. Covalent Modification of Cellular Proteins

From the early works on the antiproliferative and antiviral effects of cyPG it was realized that the cyclopentenone structure was a key structural determinant for biological activity. Interestingly, the compound cyclopentenone, a simple molecule mimicking the cyclopentenone ring was found to reproduce some of the effects of cyPG [32, 61]. Indeed, the ability of cyPG to covalently modify thiol groups is an important mechanism of action of these compounds. Studies analysing the contribution of this mechanism to the biological effects of cyPG suggest that covalent protein modification is sufficient to elicit anti-inflammatory actions. On one hand, the analog of 15d-PGJ_2, 9,10-dihydro-15d-PGJ_2, which lacks the endocyclic double bond but is a potent PPARγ agonist, shows greatly reduced ability to inhibit the induction of COX-2 or iNOS by cytokines in cellular models. This analog also failed to mimic other 15d-PGJ_2 actions, including induction of oxidative stress, reduction of cell viability [62], inhibition of CD95 ligand gene expression in T lymphocytes [63] or COX-2 induction in human breast cancer cells [64]. In contrast, simple cyclic compounds with α,β unsaturated carbonyls, such as cyclopentenone and cyclohexenone, which are inert as PPAR agonists, display anti-nflammatory effects [32]. Similar conclusions can be drawn from the use of biotinylated cyPG. Biotinylated analogs of cyPG present several interesting features. They are cell permeable and biologically active. In various studies, we and other authors have confirmed that, in spite of the presence of the bulky biotin moiety, biotinylated cyPG mimic many of the effects of their parent PG, including disruption of the vimentin cytoskeleton, inhibition of iNOS induction by cytokines or induction of permeability transition in mitochondria [23, 65]. Interestingly, biotinylated cyPG appear to be inert as PPAR agonists. PGA_1-B fails to activate a PPRE transfected in cells [66], whereas 15d-PGJ_2-B binds to PPARγ with a ten times lower affinity than 15d-PGJ_2 and fails to elicit the active conformation of the receptor (Zorrilla et al., unpublished results). Analysis of the structure of PPARγ provides a hypothesis for this lack of activity. The carboxyl group of 15d-PGJ_2 establishes interactions with defined residues in the ligand binding domain of PPARγ, and these interactions are important to elicit the productive conformational change. The presence of the biotin moiety would preclude these interactions. Therefore, biotinylated analogs of cyPG could serve as valuable probes to identify PPAR-independent effects of cyPG. A summary of the observations obtained in rat mesangial cells through the use of various compounds differing in their ability to activate PPAR and to covalently modify proteins is shown in Figure 3. This illustrates the correlation between the induction of covalent modification and the anti-inflammatory potential of several compounds.

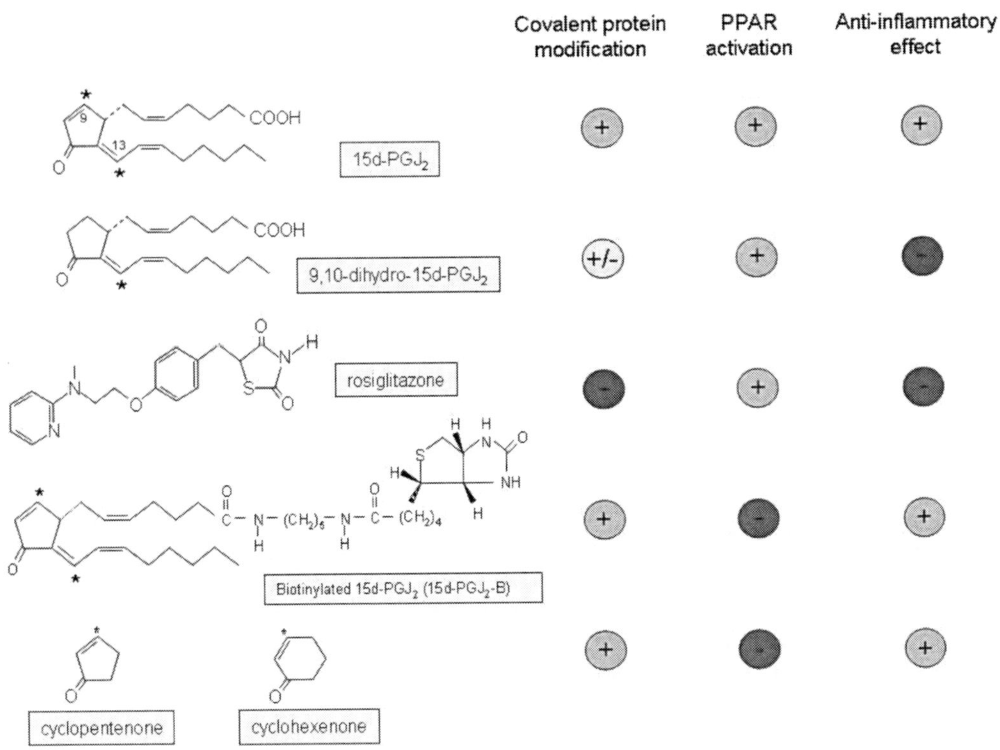

Figure 3. Correlation between protein thiol modification, PPAR agonist activity and anti-inflammatory effects of various compounds in mesangial cells. A summary of the effects observed in various works is presented [32] (Zorrilla et al., unpublished observations). Electrophilic carbons are marked by asterisks.

Since the early studies with cyPG it was recognized that they bound irreversibly to intracellular structures [67]. Both radioactive cyPG and biotinylated analogs have been used to explore the nature of the interaction of cyPG with cellular proteins. Narumiya and coworkers used radioactively labelled cyPG to show the accumulation of cyG in cells and the distribution between different cellular compartments. In later studies Sepharose-bound thiols were used as a mimic for protein thiols. This strategy showed that whereas adducts of cyPG and soluble thiols were reversible, adducts of cyPG and immobilized thiols were stable and were dissociated only after alkali treatment [68]. More recently, incubation of cells with radioactive arachidonic acid has allowed the detection of the incorporation of radioactivity into proteins [69], which has been attributed to the formation and addition of 15d-PGJ_2. Finally, biotinylated analogs of cyPG have allowed the detection of cyPG-modified polypeptides by non-radioactive methods [39, 70].

5. Identification of Protein Targets for Covalent Modification by cyPG

The identification of the proteins that are covalently modified by cyPG is important to define the mechanisms of cyPG action at a molecular level. The fact that the modification of proteins by cyPG appears to be stable under strong denaturing conditions together with the use of cyPG biotinylated analogs allow the detection of cyPG-protein adducts after SDS-PAGE analysis. This makes cyPG-protein adducts amenable to proteomic studies.

Protein	CyPG	Residue modified	Functional consequence	Reference
IKKα and β	15d-PGJ_2 PGA_1	Cys179	Inhibition	Rossi et al. 2000 Castrillo et al. 2000
NF-κB p65 subunit	15d-PGJ_2	Cys38	Inhibition	Strauss et al. 2000
NF-κB p50 subunit	15d-PGJ_2	Cys62	Inhibition	Cernuda-Morollón et al. 2001
Proteasome	15d-PGJ_2	Not identified	Inhibition	Shibata et al. 2003
H-Ras	15d-PGJ_2 PGA_1	Cys184 or Cys118	Activation	Oliva et al. 2003 Renedo et al., 2007
Thiorodoxin	15d-PGJ_2	Cys35 and Cys69	Inhibition	Shibata et al. 2003
Thioredoxin reductase	PGA_1	Not identified	Inhibition	Moos et al. 2003
Keap1	15d-PGJ_2	Several	Inhibition	Levonen et al. 2004 Kobahashi et al. 2009
c-Jun	15d-PGJ_2	Cys269	Inhibition	Pérez-Sala et al. 2003
c-Fos	15d-PGJ_2	Not identified	Inhibition	Pérez-Sala et al. 2003
Actin	15d-PGJ_2	Cys374	Structural alteration	Gayarre et al., 2006 Aldini et al., 2007
ER-α	15d-PGJ_2	Cys227 and Cys240	Inhibition	Kim et al., 2007
Vimentin	15d-PGJ_2 PGA_1	Cys328	Filament collapse	Stamatakis et al., 2006 Gharbi et al., 2007
Tubulin	15d-PGJ_2 PGA_1	Several	Structural alteration	Stamatakis et al., 2006 Gharbi et al., 2007 Cocca et al., 2009
TRPA1	15d-PGJ_2	Cys421 and Cys621	Activation	Takahashi et al., 2008
eIF-4A1	15d-PGJ_2 PGA_1	Cys264	Inhibition	Gharbi et al., 2007 Aldini et al., 2007 Kim et al., 2007

Figure 4. Summary of several protein targets for covalent modification by cyPG.

The first known protein for which the covalent binding of cyPG could be demonstrated was the NF-κB subunit p50 [39, 71]. During the last years, many targets for covalent modification by cyPG have been identified, mainly through the combination of biochemical and proteomic approaches [23, 25, 65]. These studies have allowed the identification of proteins involved in protein synthesis and degradation [72, 73], in cytoskeletal regulation, in carbohydrate metabolism and cellular redox regulation, among other essential processes. The functional consequences of modification by cyPG can be diverse. An inhibition of protein function has come forward in the case of several transcription factors, including NF-κB, AP-1, and the oestrogen receptor, for which modification of critical residues in the DNA binding domain abolishes DNA binding [37, 39, 74]. Modification of the IκB kinase, IKK, results in the inhibition of phosphorylation and subsequent degradation of IκB, thus contributing to the inhibition of the NF-κB activation pathway in some systems [75, 76]. A 15d-PGJ_2-proteasome adduct has been identified by using biotinylated 15d-PGJ_2, providing evidence for the direct modification of proteasome components by cyPG, which leads to inhibition of proteasomal degradation and may result in accumulation of misfolded proteins [77]. The modification of both thioredoxin and thioredoxin reductase by cyPG [62, 78] result in the inactivation of these proteins. This may have important implications for the regulation of cellular redox status and the regeneration of reduced thiol groups in proteins to their reduced state. Covalent protein modification appears to be also at the basis of the antiviral effects of cyPG. Besides the modification and inactivation of NF-κB, the ability of 15d-PGJ_2 to covalently modify and inhibit the HIV transactivating Tat protein has been recently reported, this having an inhibitory effect on HIV transcriptional elongation and replication [16]. In some cases, protein covalent modification results in activation of protein function. Covalent modification of PPARγ is associated with activation of the receptor, as outlined above [47]. cyPG also bring about the activation of transcription factor Nrf-2, a critical factor regulating the expression of genes involved in cellular defence and antioxidant response [79], through indirect mechanisms. In this case, the stimulatory effect rises form the modification of the negative regulator of this factor, known as Keap1 [80]. As a result of Keap1 modification, ubiquitination and degradation of Nrf2 is inhibited [81, 82]. Nuclear accumulation of Nrf2 leads to an increase in the expression of genes under the control of this transcription factor, including heme oxygenase 1 and γ-glutamilcysteine ligase [83]. Induction of these enzymes may play an important role in the defence of cells against endogenous or exogenous electrophiles. Modification of the signalling proteins of the Ras family, H-, N-, and K-Ras by cyPG also correlates with activation of Ras-mediated pathways [84]. Although the molecular mechanisms for this effect have not been elucidated, binding of cyPG occur either at the cysteine residue located in the GTP-binding site of these proteins or at the C-terminus of the protein in a Ras protein and cyPG-dependent manner (see below). Therefore, interference of the cyPG with GTP/GDP cycling or with the subcellular localization of the proteins can be envisaged as potential mechanisms for cyPG regulation of Ras proteins activity. A summary of several of the protein targets for covalent modification by cyPG is presented in Figure 4.

6. Selectivity of the Modification of Proteins by cyPG

From the wide reactivity of cyPG and the growing number of cyPG protein targets being identified it could be inferred that protein modification by cyPG is a non-specific process. However, our studies using cyPG with different structure have shown that protein modification by cyPG displays selectivity. First, the pattern of cyPG-modified proteins is different from that of total protein staining and of the pattern given by a general cysteine reagent such as biotinylated iodoacetamide [32]. Moreover, prostaglandins with different structures target different subsets of cellular proteins [85]. Several factors contribute to this selectivity including the structures of both, the cyPG and the protein. This selectivity is both inter and intra-molecular, that is, electrophilic lipids with diverse structure can selectively modify different cellular proteins or even different cysteine residues within the same protein [84]. It is clear that cyPG do not form adducts with all accessible cysteine residues in a given protein, but with specific residues, and that the residue(s) modified may be different depending on the prostanoid structure. Ras proteins provide a clear example of this. Single enone prostaglandins can bind H-Ras proteins at several cysteine residues, including the cysteine present in the GTP-binding site (C118) and the cysteines located in the C-terminal hypervariable domain (C181 and C184), which are sites for palmitoylation in cells. In contrast, cyPG with dienone structure, like 15d-PGJ_2 and Δ^{12}-PGJ_2 bind preferentially to the vicinal cysteines present in the C-terminus of H-Ras proteins (C181 and C184) [84, 86]. Therefore, this differential modification may have important consequences for the selectivity of the activation of Ras-dependent pathways by cyPG. The three Ras proteins, H, N and K-Ras possess a cysteine residue in the GTP binding site, but only H-Ras possesses two cysteine residues in the C-terminal domain. Consistent with the above observations, PGA_1, a single enone cyPG, is able to bind and activate all three Ras proteins, whereas dienone cyPG, like 15d-PGJ_2, selectively bind and activate H-Ras [84]. The structural bases for these differences are multiple. Dienone cyPG possess two electrophilic carbons and can suffer the simultaneous attack of two nucleophilic residues. In fact, dienone cyPG can act as bifunctional cysteine reagents and induce the crosslinking of cysteines in a proximity of approximately 3-6 Å, including the cysteines present in the flexible C-terminal domain of H-Ras, giving rise to an intra-molecular cross-linking, or of cysteines present in two different proteins, like the cysteines present the DNA domain of each of the c-Jun monomers forming a c-jun homodimer, leading in this case to an inter-molecular crosslinking [37].

The reactivity of the cyPG is also an important factor that influences the nature of the proteins modified in cells and in vitro. In a series of important works, Suzuki et al. explored the differential reactivity of several cyPG [68]. They found that whereas single enone cyPG reacted more slowly with GSH, the GSH-cyPG adducts formed were more stable. This could result in less cyPG being available to modify proteins in cells. In contrast, dienone cyPG reacted with GSH more rapidly but this adduct was less stable, which would favour the availability of cyPG for protein modification. Results obtained in our laboratory support this interpretation. We observed that 15d-PGJ_2-B was about 10-times more efficient than PGA_1-B at forming protein adducts when added to intact cells. These differences were cancelled when adding the cyPG to cell free lysates, but could be reproduced by varying the concentration of GSH present in the lysates. Moreover, we could vary the intensity of incorporation of biotinylated cyPG into proteins in intact cells by modulating the synthesis of GSH [85]. Interestingly, the influence of GSH was stronger in the case of PGA_1-B incorporation into proteins than in the case of the incorporation of 15d-PGJ_2-B. Thus, GSH availability inversely correlated with the extent of protein modification by PGA_1-B, whereas it affected incorporation of 15d-PGJ_2 to a lesser extent. More importantly, the presence of GSH differentially affected the incorporation of cyPG into defined polypeptides, thus modulating the selectivity of protein modification and potentially the specificity of cyPG action [85]. We have recently explored the basis for this effect by means of in vitro assays using recombinant proteins [87]. cyPG with different structure modified various recombinant proteins in vitro with different efficiency. Thus, 15d-PGJ_2-B was found to bind more efficiently than PGA_1-B to recombinant p50, whereas both cyPG bound to recombinant GST with similar potency. This could be related to the three-dimensional structure of the protein favouring the interaction with one of the cyPG depending on the spatial conformation of the side chains. Interestingly, the presence of GSH also differentially affected the modification of recombinant proteins by diverse cyPG in vitro. In particular, GSH improved cyPG incorporation into p50 but reduced the modification of GST [87]. The molecular basis for this effect can, in turn, be multiple and be related to the ability of GSH to induce changes in protein conformation or to maintain oxidation-prone thiol groups in a reduced state.

In order to obtain insight into the mechanisms and consequences of the differential modification we have recently employed proteomic approaches. This has yielded the identification of several proteins that are preferentially modified by PGA_1-B over 15d-PGJ_2-B. The results on the selectivity of protein modification by cyPG open the way for the design of compounds that may selectively target certain proteins, a first step for drug development. Taking into account the importance of structural factors for the selectivity of protein modification it would be obvious to predict that the proteins modified by biotinylated cyPG would not be necessarily the same that those modified by natural cyPG. Therefore, in order for the proteomic studies using biotinylated cyPG to yield biologically relevant data it is important that biotinylated cyPG elicit the desired biological effects. This increases the probability that the targets identified may be involved in the biological effects under study.

7. Active Concentrations of cyPG

Whereas the pharmacological effects of cyPG are well established, the pathophysiological relevance of these compounds has been the subject of intense controversy. For cyPG to play a physiologically relevant role and be considered as true lipid mediators it is important that they are generated in biological systems at concentrations high enough to elicit biological effects.

Most cellular effects of cyPG have been observed with micromolar concentrations of these compounds. The need to use relatively high concentrations may derive from the high reactivity of these compounds, by which a substantial proportion of them is lost by binding to serum proteins [83]. This reactivity may also be at the basis of the difficulties to accurately assess the concentrations of these compounds in biological systems, and may lead to an underestimation of the real concentrations of cyPG [30]. The cyPG 15d-PGJ_2 has been detected in biological samples by immunological methods. This has allowed imaging of 15d-

PGJ_2 accumulation in foamy macrophages of human atherosclerotic plaques [14] and and in spinal cord sections of patients with sporadic amyotrophic lateral sclerosis [88]. Quantitation of 15d-PGJ_2 has been attempted by immunoassay, although this method lacked full specificity. Concentrations of 15d-PGJ_2 measured through this assay in inflammatory exudates and cell supernatants fell in the picomolar or nanomolar range [37, 89]. Recently, Rajakariar et al. demonstrated bona-fide synthesis of 15d-PGJ_2, which could be detected at concentrations up to 5 ng/ml by means of liquid chromatography–tandem MS (LC-MS/MS), in inflammatory exudates of wild type but not of hematopoietic PGD_2 synthase knockout animals [36]. In an important series of works these authors have proposed a role for endogenous 15d-PGJ_2 both in the onset and in the resolution of inflammation [36, 89]. On the other hand, an increasing number of works is reporting biological effects of cyPG which may be observed with nanomolar concentrations of the compounds, close to those measured in vivo. Nanomolar concentrations of 15d-PGJ_2 induce eosinophil chemotaxis [90] and may inhibit IL-1β-induced PGE_2 production by amnion cells by mechanisms dependent on PPAR [91], among other effects.

Interestingly, a recent work by Darley-Usmar and co-workers has shown that exposure of cells to repeated additions of nanomolar concentrations of cyPG elicits effects equivalent to those of higher concentrations used in a single dose [83]. This has been proposed to be due to the accumulation of protein-cyPG adducts over time, which has been observed in several experimental systems [83, 92]. Therefore, these observations may contribute to reconcile the discrepancies between the concentrations of cyPG measured in various experimental systems (nanomolar) and the requirement for micromolar concentrations of these compounds to elicit biological effects when used in a single addition, thus suggesting that many of the effects described in cellular models may reflect biologically relevant phenomena.

8. Stability of cyPG-Protein Adducts

Another important factor for the putative role of cyPG as signaling mediators is the potential reversibility of their effects. It is generally assumed that covalent protein modification by cyPG is an irreversible process, and that cyPG-protein adducts can only be reversed by harsh biochemical conditions such as alkali treatment. However, several recent evidences call on other possibilities for cyPG-protein interaction. In the case of PPARγ, covalent modification by 15d-PGJ_2 has been shown to occur. However, pre-covalent complexes of PPAR and 15d-PGJ_2 or other endogenous fatty acid derivatives have also been evidenced [49], in what has been called a dock and lock mechanism [93]. This could occur in other proteins, with the detected covalently modified proteins being only a proportion of the cyPG-interacting protein. An example of this may be provided by GST enzymes, for which both reversible and irreversible mechanisms of inhibition by cyPG have been reported [92, 94]. Interestingly, the inhibition of GSTP1-1 by PGA_2 can be relieved by excess GSH, thus suggesting the possibility that under certain conditions retro-Michael reactions occur [95]. It has been suggested that the stability of cyPG-protein adducts may depend on the microenvironment of the modified cysteine residue [96]. This could also influence which proteins undergo stable modification. Recently, using biotin-tagged elecrophiles to model protein addition and biological effects of endogenous electrophiles it was observed that adducts of different electrophiles displayed different stability in cells. Moreover, this study suggested the involvement of metabolic processes in adduct removal [97]. This raises the exciting possibility that the repair of electrophile-protein adducts may be regulated enzymatically, and if so, identification and characterization of the enzymes responsible will be an interesting field of research.

9. THERAPEUTIC POTENTIAL OF CYPG

Since the unveiling of the anti-viral and anti-proliferative effects of cyPG, a potential therapeutic use of these compounds has been envisaged. The beneficial effects of these compounds have been explored in a variety of animal models of inflammation or tissue injury.

Micromolar concentrations of 15d-PGJ_2 have shown protective effects against ischemic acute renal failure in rats, an effect associated with the inhibition of NF-κB [98]. This cyPG also displayed beneficial effects in systemic inflammation and multiple-organ failure caused by endotoxic shock [99, 100]. Inhibitory actions of 15d-PGJ_2 on NF-κB and AP-1 [99] or activation of PPARγ [100] have been proposed as potential mechanisms for these effects. Moreover, 15d-PGJ_2 has been shown to induce sinoviocyte apoptosis, which may contribute to the suppression of adjuvant-induced arthritis in rats [101]. 15d-PGJ_2 has been found to exert protective effects in several models of gastrointestinal disease or damage, including ischemia/reperfusion injury of the gut [102], where a role for PPARγ activation has been proposed, or in colon injury caused by dinitrobenzene sulfonic acid [103]. In this model, 15d-PGJ_2 reduced the production of pro-inflammatory cytokines, iNOS expression and NF-κB activation. CyPG have been also shown to display protective effects in the cardiovascular system. PGJ_2 reduces restenosis after balloon angioplasty in rats. Interestingly, this effect is associated with inhibition of NF-κB and could be mimicked by the compound 2-cyclopentenone, which suggests the involvement of protein modification in the protective effects [104]. Also, a preconditioning of human smooth muscle cells by treatment with cyPG protects against toxic effects of oxidized low-density lipoprotein [105]. The beneficial effects of cyPG and other PPAR agonists on cardiovascular function have led to assay the effects of an endothelium-directed cyPG-liposome formulation. This preparation, known as LipoCardium, has been found to drastically reduce vascular lesions associated with atherosclerosis and improve survival in mice [106].

CyPG have been shown to exert anti-proliferative and/or pro-apoptotic actions in numerous studies using tumor cells in culture. Several studies have evidenced anti-tumoral effects of cyPG in various in vivo models [9, 19]. Moreover, cyPG inhibit proliferation and/or induce apoptosis in cell lines that are resistant to other anti-tumoral agents, which raises the possibility of using them in combination with other agents to increase their efficacy or potency [107, 108]. In addition, current efforts aim at the identification or development of cyPG analogs with improved anti-inflammatory or anti-proliferative potency (Figure 5). Nevertheless, establishing an anti-tumoral role for cyPG in vivo requires further assessment. Recent evidence indicates that the effects of 15d-PGJ_2 may be highly dependent on the experimental model and design. Depending on the experimental conditions, this cyPG can exert opposite effects on angiogenesis [109], a process critical for tumor growth. Moreover, a potentiation of tumor formation by 15d-PGJ_2 has been reported in a model of skin carcinogenesis [110]. Therefore, elucidation of the factors influencing the multifaceted role of cyPG in tumor biology will be of primary importance.

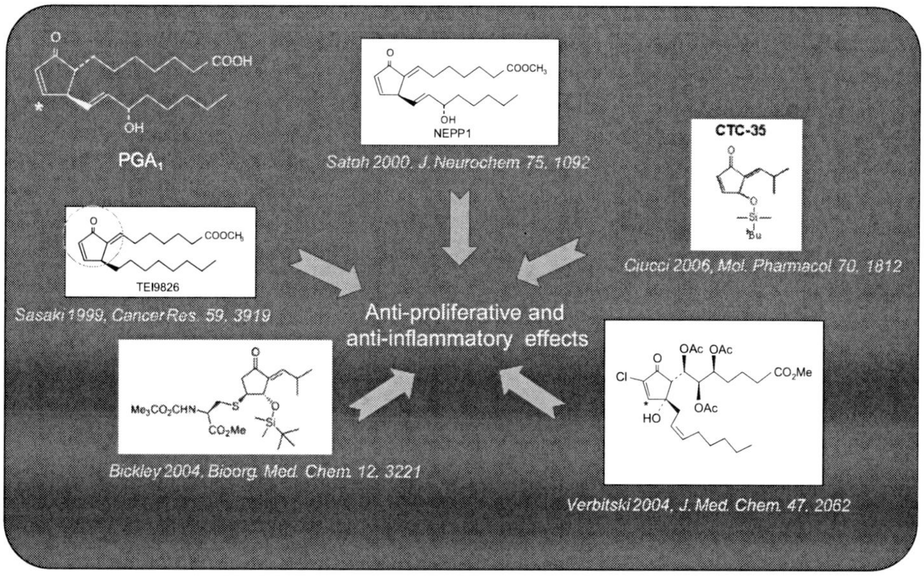

Figure 5. Several compounds bearing a PGA_1-like reactive moiety that have been studied as potential anti-inflammatory or anti-proliferative agents

The effects of cyPG observed in the nervous system are very complex. Prostaglandins of the A series have been reported to induce both neuroprotective and neurotoxic effects, the molecular basis for which is not well understood [111-113]. In a recent work we have identified several proteins of the serine biosynthetic pathway as potential targets for modification by PGA_1, namely phosphoserine aminotransferase and 3-phosphoglycerate dehydrogenase. L-serine is available from dietary sources; however, it can be also synthesized in cells by a pathway which comprises three enzymatic steps [114], two of which are catalyzed by the enzymes identified as potential PGA_1 targets. This pathway is essential for an adequate supply of serine to the brain, since permeability of the blood brain barrier to plasma serine is limited. Interestingly, serine can play a role in the inflammatory response in microglia and behave as an excitotoxic factor. Thus, the identification of serine biosynthesis enzymes as potential targets for PGA_1 may provide novel working hypothesis for the neuronal effects of cyPG. In addition, protective effects of 15d-PGJ_2 on various neurodegenerative or neurological diseases have been reported. 15d-PGJ_2 ameliorated experimental autoimmune encephalomyelitis, which constitutes an animal model for the human disease multiple sclerosis [115], and it has been shown to inhibit beta-amyloid stimulated pro-inflammatory responses and neurotoxicity [116]. However, several recent evidences also call attention on the potential adverse effects of cyPG. PGJ_2 has been reported to elicit abnormal protein aggregation leading to symptoms similar to those of neurodegenerative diseases in mice [117]. This may be due to the inhibition of proteasome activity, which would result in the accumulation of damaged proteins, or to the fact that some proteins carrying these modifications could be more resistant to degradation and accumulate in cells. Both anti-nociceptive and nociceptive responses elicited by 15d-PGJ_2 have been reported [118, 119]. Interestingly, the nociceptive effect appears to be mediated by activation of the ion channel TPRA, a member of the Transient Receptor Potential ion channel superfamily, by direct modification of the channel cysteine residues [120]. Therefore, further experimentation is needed to evaluate the potential beneficial and adverse effects of 15d-PGJ_2 on the nervous system. In this context, detailed proteomic identification of cyPG targets may contribute to clarify these factors.

10. Conclusion

CyPG have received considerable attention recently due to their potential as anti-inflammatory or anti-tumoral agents. The number of works addressing the effects of cyPG in cellular models has increased exponentially. In parallel, many studies have addressed the molecular mechanisms underlying the various effects observed. The interest of these studies is multiple. On one hand it is becoming clear that cyPG act as true lipid mediators in situations of inflammation. Therefore, elucidation of their pathophysiological role will contribute to a better knowledge of the regulation of inflammation and may aid in identifying novel targets for therapeutic intervention. Another important aspect of these studies is that many of the novel concepts that are arising from the study of cyPG may be applicable to other agents acting through similar mechanisms. Many synthetic and natural compounds may covalently bind to cysteine residues in proteins. Moreover, many of these compounds are considered potential therapeutic agents due to their anti-proliferative effect in cancer cell lines or in vivo. Future studies will need to address the efficacy and specificity of cyPG and related compounds in vivo, and perform structure-activity studies in order to find more selective and potent compounds. In parallel, proteomic studies identifying the protein targets for covalent modification may aid in the understanding of the beneficial effects and of potential adverse effects of these compounds. Therefore, coming years will probably see an interesting development of the cyPG field.

Acknowledgements: The author would like to acknowledge the contribution of students and postdoctoral fellows that have joined the laboratory of "Posttranslational modification of proteins" at the CIB, without whom this work would have not been possible: B. Garzón, B. Díez-Dacal, C. Oeste, M.I. Avellano, and Drs. E. Cernuda-Morollón, K. Stamatakis, F.J. Sánchez-Gómez, J. Gayarre, and M. Renedo. The work of many collaborators: Dr. F.J. Cañada, J.M. Rojas. L. Boscá, O. Llorca, M. Pajares, is also most gratefully acknowledged. Work in the laboratory has been supported by grants from Ministerio de Ciencia e Innovación SAF2006-03489, SAF2009-11642, and RETIC RIRAAF, RD07/0064/0007 from ISCIII.

References

[1] Funk, C. D. (2001) Prostaglandins and leukotrienes: advances in eicosanoid biology, *Science.*, *294*, 1871-1875.

[2] Norel, X. (2007). Prostanoid receptors in the human vascular wall, *ScientificWorldJournal.*, *7*, 1359-1374.

[3] Foudi, N., Kotelevets, L., Louedec, L., Leseche, G., Henin, D., Chastre, E. & Norel, X. (2008). Vasorelaxation induced by prostaglandin E2 in human pulmonary vein: role of the EP4 receptor subtype, *Br J Pharmacol., 154*, 1631-1639.

[4] Ziboh, V. A., Miller, C. C. & Cho, Y. (2000). Metabolism of polyunsaturated fatty acids by skin epidermal enzymes: generation of antiinflammatory and antiproliferative metabolites, *Am J Clin Nutr., 71*, 361S-366S.

[5] Gao, L., Zackert, W. E., Hasford, J. J., Danekis, M. E., Milne, G. L., Remmert, C., Reese, J., Yin, H., Tai, H. H., Dey, S. K., Porter, N. A. & Morrow, J. D. (2003). Formation of prostaglandins E2 and D2 via the isoprostane pathway: a mechanism for the generation of bioactive prostaglandins independent of cyclooxygenase., *J. Biol. Chem., 278*, 28479-28489.

[6] Zusman, R. M., Caldwell, B. V., Speroff, L. & Behrman, H. R. (1972). Radioimmunoassay of the A prostaglandins, *Prostaglandins., 2*, 41-53.

[7] Jaffe, B. M., Behrman, H. R. & Parker, C. W. (1973). Radioimmunoassay measurement of prostaglandins E, A, and F in human plasma, *J Clin Invest., 52*, 398-405.

[8] Santoro, M. G., Benedetto, A., Carruba, G., Garaci, E. & Jaffe, B. M. (1980). Prostaglandin A compounds as antiviral agents, *Science., 209*, 1032-1034.

[9] Kato, T., Fukushima, M., Kurozumi, S. & Noyori, R. (1986). Antitumor activity of delta 7-prostaglandin A1 and delta 12- prostaglandin J2 in vitro and in vivo, *Cancer Res., 46*, 3538-3542.

[10] Fukushima, M., Kato, T., Ota, K., Arai, Y., Narumiya, S. & Hayaishi, O. (1982). 9-deoxy-delta 9-prostaglandin D2, a prostaglandin D2 derivative with potent antineoplastic and weak smooth muscle-contracting activities, *Biochem Biophys Res Commun., 109*, 626-633.

[11] Fitzpatrick, F. A. & Wynalda, M. A. (1983). Albumin-catalyzed metabolism of prostaglandin D2. Identification of products formed in vitro, *J. Biol. Chem., 258*, 11713-11718.

[12] Kikawa, Y., Narumiya, S., Fukushima, M., Wakatsuka, H. & Hayaishi, O. (1984). 9-Deoxy-delta 9, delta 12-13,14-dihydroprostaglandin D2, a metabolite of prostaglandin D2 formed in human plasma., *Proc Natl Acad Sci*, U S A. 81, 1317-1321.

[13] Hirata, Y., Hayashi, H., Ito, S., Kikawa, Y., Ishibashi, M., Sudo, M., Miyazaki, H., Fukushima, M., Narumiya, S. & Hayaishi, O. (1988). Occurrence of 9-deoxy-delta 9,delta 12-13,14-dihydroprostaglandin D2 in human urine, *J Biol Chem., 263*, 16619-16625.

[14] Shibata, T., Kondo, M., Osawa, T., Shibata, N., Kobayashi, M. & Uchida, K. (2002). 15-Deoxy-Delta 12,14-prostaglandin J2. A prostaglandin D2 metabolite generated during inflammatory processes, *J Biol Chem., 277*, 10459 10466.

[15] Santoro, M. G. (1997). Antiviral activity of cyclopentenone prostanoids, *Trends Microbiol., 5*, 276-281.

[16] Kalantari, P., Narayan, V., Henderson, A. J. & Prabhu, K. S. (2009). 15-Deoxy-Delta12,14-prostaglandin J2 inhibits HIV-1 transactivating protein, Tat, through covalent modification, *Faseb J., 23*, 2366-2373.

[17] Fukushima, M., Kato, T., Ueda, R., Ota, K., Narumiya, S. & Hayaishi, O. (1982). Prostaglandin D2, a potential antineoplastic agent, *Biochem Biophys Res Commun., 105*, 956-964.

[18] Bregman, M. D., Funk, C. & Fukushima, M. (1986). Inhibition of human melanoma

growth by In: A, D. prostaglandin, & analogues, *J. Cancer Res.*, *46*, 2740-2744.

[19] Sasaki, H., Niimi, S., Akiyama, M., Tanaka, T., Hazato, A., Kurozumi, S., Fukushima, S. & Fukushima, M. (1999). Antitumor activity of 13,14-dihydro-15-deoxy-delta7-prostaglandin-A1-methyl ester integrated into lipid microspheres against human ovarian carcinoma cells resistant to cisplatin in vivo., *Cancer Res.*, *59*, 3919-3922.

[20] Tanikawa, M., Yamada, K., Tominaga, K., Morisaki, H., Kaneko, Y., Ikeda, K., Suzuki, M., Kiho, T., Tomokiyo, K., Furuta, K., Noyori, R. & Nakanishi, M. (1998). Potent prostaglandin A1 analogs that suppress tumor cell growth through induction of p21 and reduction of cyclin E, *J Biol Chem.*, *273*, 18522-18527.

[21] Campo, P. A., Das, S., Hsiang, C. H., Bui, T., Samuel, C. E. & Straus, D. S. (2002). Translational regulation of cyclin D1 by 15-deoxy-delta(12,14)-prostaglandin J(2), *Cell Growth Differ.*, *13*, 409-420.

[22] Kamagata, C., Tsuji, N., Moriai, M., Kobayashi, D. & Watanabe, N. (2007). 15-deoxy-Delta(12,14)-prostaglandin J2 inhibits G2-M phase progression in human breast cancer cells via the down-regulation of cyclin B1 and survivin expression, *Breast Cancer Res Treat.*, *102*, 263-273.

[23] Stamatakis, K., Sánchez-Gómez, F. J. & Pérez-Sala, D. (2006). Identification of novel protein targets for modification by 15-deoxy-$D^{12,14}$-prostaglandin J_2 in mesangial cells reveals multiple interactions with the cytoskeleton., *J Am Soc Nephrol.*, *17*, 89-98.

[24] Gayarre, J., Sánchez, D., Sánchez-Gómez, F. J., Terrón, C., Llorca, O. & Pérez-Sala, D. (2006). Addition of electrophilic lipids to actin alters filament structure, *Biochem Biophys Res Commun.*, *349*, 1387-1393.

[25] Aldini, G., Carini, M., Vistoli, G., Shibata, T., Kusano, Y., Gamberoni, L., Dalle-Donne, I., Milzani, A. & Uchida, K. (2007). Identification of Actin as a 15-Deoxy-Delta(12,14)-prostaglandin J(2) Target in Neuroblastoma Cells: Mass Spectrometric, Computational, and Functional Approaches To Investigate the Effect on Cytoskeletal Derangement, *Biochemistry.*, *46*, 2707-2718.

[26] Cocca, C., Dorado, J., Calvo, E., Lopez, J. A., Santos, A. & Perez-Castillo, A. (2009). 15-Deoxi-Delta(12,14)-prostaglandin J2 is a tubulin-binding agent that destabilizes microtubules and induces mitotic arrest, *Biochem Pharmacol.*, *78*, 1330-1339.

[27] Martinez, B., Perez-Castillo, A. & Santos, A. (2005). The mitochondrial respiratory complex I is a target for 15-deoxy-delta12,14-prostaglandin J2 action, *J Lipid Res.*, *46*, 736-743.

[28] Landar, A., Zmijewski, J. W., Dickinson, D. A., Legoffe, C., Johnson, M., Milne, G., Zanoni, G., Vidari, G., Morrow, J. & Darley-Usmar, V. M. (2006). The interaction of electrophilic lipid oxidation products with mitochondria in endothelial cells and the formation of reactive oxygen species, *Am J Physiol Heart Circ Physiol.*, *290*, H1777-H1787.

[29] Shin, S. W., Seo, C. Y., Han, H., Han, J. Y., Jeong, J. S., Kwak, J. Y. & Park, J. I. (2009). 15d-PGJ2 induces apoptosis by reactive oxygen species-mediated inactivation of Akt in leukemia and colorectal cancer cells and shows in vivo antitumor activity, *Clin Cancer Res.*, *15*, 5414-5425.

[30] Uchida, K. & Shibata, T. (2008). 15-Deoxy-Delta(12,14)-prostaglandin J2: an electrophilic trigger of cellular responses, *Chem Res Toxicol.*, *21*, 138-144.

[31] Ricote, M., Li, A. C., Willson, T. M., Kelly, C. J. & Glass, C. K. (1998). The peroxisome proliferator-activated receptor-γ is a negative regulator of macrophage

activation, *Nature.*, *391*, 79-82.

[32] Sánchez-Gómez, F. J., Cernuda-Morollón, E., Stamatakis, K. & Pérez-Sala, D. (2004). Protein thiol modification by 15-deoxy-$\Delta^{12,14}$-prostaglandin J_2 addition in mesangial cells: role in the inhibition of pro-inflammatory genes, *Mol. Pharmacol.*, *66*, 1349-1358.

[33] Jiang, C., Ting, A. T. & Seed, B. (1998). PPAR-gamma agonists inhibit production of monocyte inflammatory cytokines, *Nature.*, *391*, 82-86.

[34] Gilroy, D. W., Lawrence, T., Perretti, M. & Rossi, A. G. (2004). Inflammatory resolution: new opportunities for drug discovery, *Nat Rev Drug Discov.*, *3*, 401-416.

[35] Hortelano, S., Castrillo, A., Alvarez, A. M. & Boscá, L. (2000). Contribution of cyclopentenone prostaglandins to the resolution of inflammation through the potentiation of apoptosis in activated macrophages, *J. Immunol.*, *165*, 6525-6531.

[36] Rajakariar, R., Hilliard, M., Lawrence, T., Trivedi, S., Colville-Nash, P., Bellingan, G., Fitzgerald, D., Yaqoob, M. M. & Gilroy, D. W. (2007). Hematopoietic prostaglandin D2 synthase controls the onset and resolution of acute inflammation through PGD2 and 15-deoxyDelta12 14 PGJ2, *Proc Natl Acad Sci.*, U S A., *104*, 20979-20984.

[37] Pérez-Sala, D., Cernuda-Morollón, E. & Cañada, F. J. (2003). Molecular basis for the inhibition of AP-1 DNA binding by 15-deoxy-$\Delta^{12,14}$-prostaglandin J_2, *J. Biol. Chem.*, *278*, 51251-51260.

[38] Straus, D. S., Pascual, G., Li, M., Welch, J. S., Ricote, M., Hsiang, C. H., Sengchanthalangsy, L. L., Ghosh, G. & Glass, C. K. (2000). 15-deoxy-delta 12,14-prostaglandin J2 inhibits multiple steps in the NF- kappa B signaling pathway, *Proc Natl Acad Sci*, U S A., *97*, 4844-4849.

[39] Cernuda-Morollón, E., Pineda-Molina, E., Cañada, F. J. & Pérez-Sala, D. (2001). 15-Deoxy-$\Delta^{12,14}$-prostaglandin J_2 inhibition of NF-κB DNA binding through covalent modification of the p50 subunit, *J. Biol. Chem.*, *276*, 35530-35536.

[40] Li, M., Pascual, G. & Glass, C. K. (2000). Peroxisome proliferator-activated receptor gamma-dependent repression of the inducible nitric oxide synthase gene, *Mol Cell Biol.*, *20*, 4699-4707.

[41] Delerive, P., Bosscher, K. D., Besnard, S., Berghe, W. V., Peter, J. M., Gonzalez, F. J., Fruchart, J.-C., Tedgui, A., Haegeman, G. & Staels, B. (1999). Peroxisome proliferator-activated receptor a negatively regulates the vascular inflammatory gene response by negative cross-talk with transcription factors NF-κB and AP-1, *J. Biol. Chem.*, *274*, 32048-32054.

[42] Delerive, P., Gervois, P., Fruchart, J. C. & Staels, B. (2000). Induction of IkappaBalpha expression as a mechanism contributing to the anti-inflammatory activities of peroxisome proliferator-activated receptor-alpha activators, *J Biol Chem.*, *275*, 36703-36707.

[43] Francois, M., Richette, P., Tsagris, L., Raymondjean, M., Fulchignoni-Lataud, M. C., Forest, C., Savouret, J. F. & Corvol, M. T. (2004). Peroxisome proliferator-activated receptor-gamma down-regulates chondrocyte matrix metalloproteinase-1 via a novel composite element, *J Biol Chem.*, *279*, 28411-28418.

[44] Levonen, A. L., Dickinson, D. A., Moellering, D. R., Mulcahy, R. T., Forman, H. J. & Darley-Usmar, V. M. (2001). Biphasic effects of 15-deoxy-delta(12,14)-prostaglandin J(2) on glutathione induction and apoptosis in human endothelial cells, *Arterioscler*

Thromb Vasc Biol., *21*, 1846-1851.

[45] Forman, B. M., Tontonoz, P., Chen, J., Brun, R. P., Spiegelman, B. M. & Evans, R. M. (1995). 15-Deoxy-delta 12, 14-prostaglandin J2 is a ligand for the adipocyte determination factor PPAR gamma, *Cell.*, *83*, 803-812.

[46] Kliewer, S. A., Lenhard, J. M., Willson, T. M., Patel, I., Morris, D. C. & Lehmann, J. M. (1995). A prostaglandin J2 metabolite binds peroxisome proliferator-activated receptor gamma and promotes adipocyte differentiation, *Cell.*, *83*, 813-819.

[47] Shiraki, T., Kamiya, N., Shiki, S., Kodama, T. S., Kakizuka, A. & Jingami, H. (2005). α,ß-Unsaturated ketone is a core moiety of natural ligands for covalent binding to peroxisome proliferator-activated receptor, *J Biol Chem.*, *280*, 14145-14153.

[48] Itoh, T., Fairall, L., Amin, K., Inaba, Y., Szanto, A., Balint, B. L., Nagy, L., Yamamoto, K. & Schwabe, J. W. (2008). Structural basis for the activation of PPARgamma by oxidized fatty acids, *Nat Struct Mol Biol.*, *15*, 924-931.

[49] Waku, T., Shiraki, T., Oyama, T., Fujimoto, Y., Maebara, K., Kamiya, N., Jingami, H. & Morikawa, K. (2009). Structural insight into PPARgamma activation through covalent modification with endogenous fatty acids, *J Mol Biol.*, *385*, 188-199.

[50] Barak, Y. & Kim, S. (2007). Genetic Manipulations of PPARs: Effects on Obesity and Metabolic Disease, *PPAR Res.*, 12781.

[51] Hevener, A. L., Olefsky, J. M., Reichart, D., Nguyen, M. T., Bandyopadyhay, G., Leung, H. Y., Watt, M. J., Benner, C., Febbraio, M. A., Nguyen, A. K., Folian, B., Subramaniam, S., Gonzalez, F. J., Glass, C. K. & Ricote, M. (2007). Macrophage PPAR gamma is required for normal skeletal muscle and hepatic insulin sensitivity and full antidiabetic effects of thiazolidinediones, *J Clin Invest.*, *117*, 1658-1669.

[52] Bishop-Bailey, D. & Hla, T. (1999). Endothelial cell apoptosis induced by the peroxisome proliferator-activated receptor (PPAR) ligand 15-deoxy-Delta12, 14-prostaglandin J2, *J Biol Chem.*, *274*, 17042-17048.

[53] Yu, K., Bayona, W., Kallen, C. B., Harding, H. P., Ravera, C. P., McMahon, G., Brown, M. & Lazar, M. A. (1995). Differential activation of peroxisome proliferator-activated receptors by eicosanoids, *J Biol Chem.*, *270*, 23975-83.

[54] Forman, B. M., Chen, J. & Evans, R. M. (1997). Hypolipidemic drugs, polyunsaturated fatty acids, and eicosanoids are ligands for peroxisome proliferator-activated receptors alpha and delta, *Proc Natl Acad Sci*, U S A., *94*, 4312-4317.

[55] Sawyer, N., Cauchon, E., Chateauneuf, A., Cruz, R. P., Nicholson, D. W., Metters, K. M., O'Neill, G. P. & Gervais, F. G. (2002). Molecular pharmacology of the human prostaglandin D2 receptor, CRTH2, Br *J Pharmacol.*, *137*, 1163-1172.

[56] Gazi, L., Gyles, S., Rose, J., Lees, S., Allan, C., Xue, L., Jassal, R., Speight, G., Gamble, V. & Pettipher, R. (2005). Delta12-prostaglandin D2 is a potent and selective CRTH2 receptor agonist and causes activation of human eosinophils and Th2 lymphocytes, *Prostaglandins Other Lipid Mediat.*, *75*, 153-167.

[57] Sandig, H., Pease, J. E. & Sabroe, I. (2007). Contrary prostaglandins: the opposing roles of PGD2 and its metabolites in leukocyte function, *J Leukoc Biol.*, *81*, 372-382.

[58] Monneret, G., Li, H., Vasilescu, J., Rokach, J. & Powell, W. S. (2002). 15-Deoxy-delta 12,14-prostaglandins D2 and J2 are potent activators of human eosinophils, *J Immunol.*, *168*, 3563-3569.

[59] Hata, A. N., Zent, R., Breyer, M. D. & Breyer, R. M. (2003). Expression and molecular pharmacology of the mouse CRTH2 receptor, *J. Pharmacol. Exp. Ther.*, *306*, 463-470.

[60] Gosset, P., Bureau, F., Angeli, V., Pichavant, M., Faveeuw, C., Tonnel, A. B. & Trottein, F. (2003). Prostaglandin D2 affects the maturation of human monocyte-derived dendritic cells: consequence on the polarization of naive Th cells, *J Immunol.*, *170*, 4943-4952.

[61] Ianaro, A., Ialenti, A., Maffia, P., Di Meglio, P., Di Rosa, M. & Santoro, M. G. (2003). Anti-inflammatory activity of 15-deoxy-delta12,14-PGJ2 and 2-cyclopenten-1-one: role of the heat shock response, *Mol. Pharmacol.*, *64*, 85-93.

[62] Shibata, T., Yamada, T., Ishii, T., Kumazawa, S., Nakamura, H., Masutani, H., Yodoi, J. & Uchida, K. (2003). Thioredoxin as a molecular target of cyclopentenone prostaglandins, *J Biol Chem.*, *278*, 26046-26054.

[63] Cippitelli, M., Fionda, C., Di Bona, D., Lupo, A., Piccoli, M., Frati, L. & Santoni, A. (2003). The cyclopentenone-type prostaglandin 15-deoxy-delta 12,14-prostaglandin J2 inhibits CD95 ligand gene expression in T lymphocytes: interference with promoter activation via peroxisome proliferator-activated receptor-gamma-independent mechanisms, *J. Immunol.*, *170*, 4578-4592.

[64] Kim, E. H., Na, H. K., Kim, D. H., Park, S. A., Kim, H. N., Song, N. Y. & Surh, Y. J. (2008). 15-Deoxy-Delta12,14-prostaglandin J2 induces COX-2 expression through Akt-driven AP-1 activation in human breast cancer cells: a potential role of ROS, *Carcinogenesis.*, *29*, 688-695.

[65] Landar, A., Shiva, S., Levonen, A. L., Oh, J. Y., Zaragoza, C., Johnson, M. S. & Darley-Usmar, V. M. (2006). Induction of the permeability transition and cytochrome c release by 15-deoxy prostaglandin J2 in mitochondria, *Biochem J.*, *394*, 185-195.

[66] Garzón, B., Gayarre, J., Gharbi, S., Díez-Dacal, B., Sánchez-Gómez, F. J., Timms, J. F. & Pérez-Sala, D. (2009). A biotinylated analog of the antiproliferative prostaglandin A1 allows assessment of PPAR-independent effects and identification of novel cellular targets for covalent modification, *Chem Biol Interact*, In press.

[67] Narumiya, S., Ohno, K., Fukushima, M. & Fujiwara, M. (1987). Site and mechanism of growth inhibition by prostaglandins. III. Distribution and binding of prostaglandin A2 and delta 12-prostaglandin J2 in nuclei, *J Pharmacol Exp Ther.*, *242*, 306-311.

[68] Suzuki, M., Mori, M., Niwa, T., Hirata, R., Furuta, K., Ishikawa, T. & Noyori, R. (1997). Chemical implications for antitumor and antiviral prostaglandins: reaction of D^7-prostaglandin A_1 and prostaglandin A_1 methyl esters with thiols, *J. Am. Chem. Soc.*, *119*, 2376-2385.

[69] Vunta, H., Davis, F., Palempalli, U. D., Bhat, D., Arner, R. J., Thompson, J. T., Peterson, D. G., Reddy, C. C. & Prabhu, K. S. (2007). The anti-inflammatory effects of selenium are mediated through 15-deoxy-Delta12,14-prostaglandin J2 in macrophages, *J Biol Chem.*, *282*, 17964-73.

[70] Parker, J. (1995). Prostaglandin A_2 protein interactions and inhibition of cellular proliferation, *Prostaglandins.*, *50*, 359-375.

[71] Cernuda-Morollón, E. & Pérez-Sala, D. (2006). Regulation of proinflammatory transcription factors by direct modification with cyclopentenone prostaglandins in Trends in DNA Research (Woods, C. R., ed) pp. 1-31, Nova Science Publishers, Inc., Hauppauge, NY.

[72] Gharbi, S., Garzón, B., Gayarre, J., Timms, J. & Pérez-Sala, D. (2007). Study of protein targets for covalent modification by the antitumoral and anti-inflammatory prostaglandin PGA_1: focus on vimentin, *J Mass Spectrom.*, *42*, 1474-1484.

[73] Kim, W. J., Kim, J. H. & Jang, S. K. (2007). Anti-inflammatory lipid mediator 15d-PGJ2 inhibits translation through inactivation of eIF4A, *Embo J.*, *26*, 5020-5032.

[74] Kim, H. J., Kim, J. Y., Meng, Z., Wang, L. H., Liu, F., Conrads, T. P., Burke, T. R., Veenstra, T. D. & Farrar, W. L. (2007). 15-deoxy-Delta12,14-prostaglandin J2 inhibits transcriptional activity of estrogen receptor-alpha via covalent modification of DNA-binding domain, *Cancer Res.*, *67*, 2595-2602.

[75] Castrillo, A., Diaz-Guerra, M. J., Hortelano, S., Martin-Sanz, P. & Bosca, L. (2000). Inhibition of IkappaB kinase and IkappaB phosphorylation by 15-deoxy- Delta(12,14)-prostaglandin J(2) in activated murine macrophages, *Mol Cell Biol.*, *20*, 1692-8.

[76] Rossi, A., Kapahi, P., Natoli, G., Takahashi, T., Chen, Y., Karin, M. & Santoro, M. G. (2000). Anti-inflammatory cyclopentenone prostaglandins are direct inhibitors of IkB kinase, *Nature.*, *403*, 103-108.

[77] Shibata, T., Yamada, T., Kondo, M., Tanahashi, N., Tanaka, K., Nakamura, H., Masutani, H., Yodoi, J. & Uchida, K. (2003). An endogenous electrophile that modulates the regulatory mechanism of protein turnover: inhibitory effects of 15-deoxy-Delta 12,14-prostaglandin J2 on proteasome, *Biochemistry.*, *42*, 13960-13968.

[78] Moos, P. J., Edes, K., Cassidy, P., Massuda, E. & Fitzpatrick, F. A. (2003). Electrophilic Prostaglandins and Lipid Aldehydes Repress Redox- sensitive Transcription Factors p53 and Hypoxia-inducible Factor by Impairing the Selenoprotein Thioredoxin Reductase, *J Biol Chem.*, *278*, 745-750.

[79] Kobayashi, M. & Yamamoto, M. (2006). Nrf2-Keap1 regulation of cellular defense mechanisms against electrophiles and reactive oxygen species, *Adv Enzyme Regul.*, *46*, 113-140.

[80] Kobayashi, M., Li, L., Iwamoto, N., Nakajima-Takagi, Y., Kaneko, H., Nakayama, Y., Eguchi, M., Wada, Y., Kumagai, Y. & Yamamoto, M. (2009). The antioxidant defense system Keap1-Nrf2 comprises a multiple sensing mechanism for responding to a wide range of chemical compounds, *Mol Cell Biol.*, *29*, 493-502.

[81] Zhang, D. D. & Hannink, M. (2003). Distinct cysteine residues in Keap1 are required for Keap1-dependent ubiquitination of Nrf2 and for stabilization of Nrf2 by chemopreventive agents and oxidative stress., *Mol Cell Biol.*, *23*, 8137-8151.

[82] Kobayashi, A., Kang, M. I., Watai, Y., Tong, K. I., Shibata, T., Uchida, K. & Yamamoto, M. (2006). Oxidative and electrophilic stresses activate Nrf2 through inhibition of ubiquitination activity of Keap1, *Mol Cell Biol.*, *26*, 221-229.

[83] Oh, J. Y., Giles, N., Landar, A. & Darley-Usmar, V. (2008). Accumulation of 15-deoxy-delta(12,14)-prostaglandin J2 adduct formation with Keap1 over time: effects on potency for intracellular antioxidant defence induction, *Biochem J.*, *411*, 297-306.

[84] Renedo, M., Gayarre, J., García-Domínguez, C. A., Pérez-Rodríguez, A., Prieto, A., Cañada, F. J., Rojas, J. M. & Pérez-Sala, D. (2007). Modification and activation of Ras proteins by electrophilic prostanoids with different structure are site-selective, Biochemistry. 46, 6607-6616.

[85] Gayarre, J., Stamatakis, K., Renedo, M. & Pérez-Sala, D. (2005). Differential selectivity of protein modification by the cyclopentenone prostaglandins PGA_1 and 15-deoxy-$\Delta^{12,14}$-PGJ_2: role of glutathione, *FEBS Lett.*, *579*, 5803-5808.

[86] Oliva, J. L., Pérez-Sala, D., Castrillo, A., Martínez, N., Cañada, F. J., Boscá, L. & Rojas, J. M. (2003). The cyclopentenone 15-deoxy-$\Delta^{12,14}$-prostaglandin J_2 binds to and

activates H-Ras, *Proc Natl Acad Sci.*, U S A., *100*, 4772-4777.

[87] Gayarre, J., Avellano, M. I., Sánchez-Gómez, F. J., Carrasco, M. J., Cañada, F. J. & Pérez-Sala, D. (2007). Modification of proteins by cyclopentenone prostaglandins is differentially modulated by GSH in vitro, *Ann N Y Acad Sci.*, *1096*, 78-85.

[88] Kondo, M., Shibata, T., Kumagai, T., Osawa, T., Shibata, N., Kobayashi, M., Sasaki, S., Iwata, M., Noguchi, N. & Uchida, K. (2002). 15-Deoxy-Delta 12,14-prostaglandin J2: The endogenous electrophile that induces neuronal apoptosis, *Proc Natl Acad Sci.*, U S A. 99, 7367-7372.

[89] Gilroy, D. W., Colville-Nash, P. R., Willis, D., Chivers, J., Paul-Clark, M. J. & Willoughby, D. A. (1999). Inducible cyclooxygenase may have anti-inflammatory properties, *Nat Med.*, *5*, 698-701.

[90] Kobayashi, Y., Ueki, S., Mahemuti, G., Chiba, T., Oyamada, H., Saito, N., Kanda, A., Kayaba, H. & Chihara, J. (2005). Physiological levels of 15-deoxy-Delta12,14-prostaglandin J2 prime eotaxin-induced chemotaxis on human eosinophils through peroxisome proliferator-activated receptor-gamma ligation, *J Immunol.*, *175*, 5744-5750.

[91] Berry, E. B., Keelan, J. A., Helliwell, R. J., Gilmour, R. S. & Mitchell, M. D. (2005). Nanomolar and micromolar effects of 15-deoxy-delta 12,14-prostaglandin J2 on amnion-derived WISH epithelial cells: differential roles of peroxisome proliferator-activated receptors gamma and delta and nuclear factor kappa B, *Mol Pharmacol.*, *68*, 169-178.

[92] Sánchez-Gómez, F. J., Gayarre, J., Avellano, M. I. & Pérez-Sala, D. (2007). Direct evidence for the covalent modification of glutathione-S-transferase P1-1 by electrophilic prostaglandins: implications for enzyme inactivation and cell survival, *Arch Biochem Biophys.*, *457*, 150-159.

[93] Shiraki, T., Kodama, T. S., Shiki, S., Nakagawa, T. & Jingami, H. (2006). Spectroscopic analyses of the binding kinetics of 15d-PGJ2 to the PPARgamma ligand-binding domain by multi-wavelength global fitting, *Biochem J.*, *393*, 749-755.

[94] Paumi, C. M., Smitherman, P. K., Townsend, A. J. & Morrow, C. S. (2004). Glutathione S-Transferases (GSTs) inhibit transcriptional activation by the peroxisomal proliferator-activated receptor g (PPARγ) ligand, 15-deoxy-$\Delta^{12.14}$prostaglandin J2 (15-d-PGJ_2), *Biochemistry.*, *43*, 2345-2352.

[95] van Iersel, M. L., Cnubben, N. H., Smink, N., Koeman, J. H. & van Bladeren, P. J. (1999). Interactions of prostaglandin A2 with the glutathione-mediated biotransformation system, *Biochem Pharmacol.*, *57*, 1383-1390.

[96] Bickley, J. F., Ciucci, A., Evans, P., Roberts, S. M., Ross, N. & Santoro, G. M. (2004). Reactions of some cyclopentenones with selected cysteine derivatives and biological activities of the product thioethers, *Bioorg Med Chem.*, *12*, 3221-3227.

[97] Lin, D., Saleh, S. & Liebler, D. C. (2008). Reversibility of covalent electrophile-protein adducts and chemical toxicity, *Chem Res Toxicol.*, *21*, 2361-2369.

[98] Chatterjee, P. K., Patel, N. S., Cuzzocrea, S., Brown, P. A., Stewart, K. N., Mota-Filipe, H., Britti, D., Eberhardt, W., Pfeilschifter, J. & Thiemermann, C. (2004). The cyclopentenone prostaglandin 15-deoxy-Delta(12,14)-prostaglandin J(2) ameliorates ischemic acute renal failure, Cardiovasc. *Res.*, *61*, 630-643.

[99] Zingarelli, B., Sheehan, M., Hake, P. W., O'Connor, M., Denenberg, A. & Cook, A. (2003). Peroxisome proliferator activator receptor-gamma ligands, 15-deoxy-

delta(12,14)-prostaglandin J2 and ciglitazone, reduce systemic inflammation in polymicrobial sepsis by modulation of signal transduction pathways, *J. Immunol., 171*, 6827-6837.

[100] Dugo, L., Collin, M., Cuzzocrea, S. & Thiemermann, C. (2004). 15d-prostaglandin J2 reduces multiple organ failure caused by wall-fragment of Gram-positive and Gram-negative bacteria, *Eur J Pharmacol., 498*, 295-301.

[101] Kawahito, Y., Kondo, M., Tsubouchi, Y., Hashiramoto, A., Bishop-Bailey, D., Inoue, K., Kohno, M., Yamada, R., Hla, T. & Sano, H. (2000). 15-deoxy-delta(12,14)-PGJ(2) induces synoviocyte apoptosis and suppresses adjuvant-induced arthritis in rats, *J Clin Invest., 106*, 189-197.

[102] Cuzzocrea, S., Pisano, B., Dugo, L., Ianaro, A., Patel, N. S., Di Paola, R., Genovese, T., Chatterjee, P. K., Di Rosa, M., Caputi, A. P. & Thiemermann, C. (2003). Rosiglitazone and 15-deoxy-Delta12,14-prostaglandin J2, ligands of the peroxisome proliferator-activated receptor-gamma (PPAR-gamma), reduce ischaemia/reperfusion injury of the gut, Br. *J. Pharmacol., 140*, 366-376.

[103] Cuzzocrea, S., Ianaro, A., Wayman, N. S., Mazzon, E., Pisano, B., Dugo, L., Serraino, I., Di Paola, R., Chatterjee, P. K., Di Rosa, M., Caputi, A. P. & Thiemermann, C. (2003). The cyclopentenone prostaglandin 15-deoxy-delta(12,14)- PGJ2 attenuates the development of colon injury caused by dinitrobenzene sulphonic acid in the rat, Br. *J. Pharmacol., 138*, 678-688.

[104] Ianaro, A., Maffia, P., Cuzzocrea, S., Mazzon, E., Santoro, M. G., Di Rosa, M. & Ialenti, A. (2003). 2-Cyclopenten-1-one and prostaglandin J2 reduce restenosis after balloon angioplasty in rats: role of NF-kappaB, *FEBS Lett., 553*, 21-27.

[105] Hamel, L., Kenney, M., Clark, K., Merkel, L. & Rojas, C. (2000). Preconditioning of human smooth muscle cells via cyclopentenone prostaglandins protects against toxic effects of oxidized low-density lipoprotein, *Biochem Biophys Res Commun., 269*, 327-330.

[106] Homem de Bittencourt, P. I., Jr., Lagranha, D. J., Maslinkiewicz, A., Senna, S. M., Tavares, A. M., Baldissera, L. P., Janner, D. R., Peralta, J. S., Bock, P. M., Gutierrez, L. L., Scola, G., Heck, T. G., Krause, M. S., Cruz, L. A., Abdalla, D. S., Lagranha, C. J., Lima, T. & Curi, R. (2007). LipoCardium: endothelium-directed cyclopentenone prostaglandin-based liposome formulation that completely reverses atherosclerotic lesions, *Atherosclerosis., 193*, 245-258.

[107] Ciucci, A., Gianferretti, P., Piva, R., Guyot, T., Snape, T. J., Roberts, S. M. & Santoro, M. G. (2006). Induction of apoptosis in estrogen receptor-negative breast cancer cells by natural and synthetic cyclopentenones: role of the IkappaB kinase/nuclear factor-kappaB pathway, *Mol Pharmacol., 70*, 1812-1821.

[108] Han, H., Shin, S. W., Seo, C. Y., Kwon, H. C., Han, J. Y., Kim, I. H., Kwak, J. Y. & Park, J. I. (2007). 15-Deoxy-delta 12,14-prostaglandin J2 (15d-PGJ 2) sensitizes human leukemic HL-60 cells to tumor necrosis factor-related apoptosis-inducing ligand (TRAIL)-induced apoptosis through Akt downregulation, *Apoptosis., 12*, 2101-2114.

[109] Kim, E. H. & Surh, Y. J. (2008). The role of 15-deoxy-delta(12,14)-prostaglandin J(2), an endogenous ligand of peroxisome proliferator-activated receptor gamma, in tumor angiogenesis, *Biochem Pharmacol., 76*, 1544-1553.

[110] Millán, O., Rico, D., Peinado, H., Zarich, N., Stamatakis, K., Pérez-Sala, D., Rojas, J. M., Cano, A. & Boscá, L. (2006). Potentiation of tumor formation by topic

administration of 15-deoxy-$\Delta^{12,14}$-prostaglandin J_2 in a model of skin carcinogenesis, *Carcinogenesis.*, *27*, 328-336.

[111] Qin, Z. H., Wang, Y., Chen, R. W., Wang, X., Ren, M., Chuang, D. M. & Chase, T. N. (2001). Prostaglandin A(1) protects striatal neurons against excitotoxic injury in rat striatum, *J Pharmacol Exp Ther.*, *297*, 78-87.

[112] Zhang, H. L., Gu, Z. L., Savitz, S. I., Han, F., Fukunaga, K. & Qin, Z. H. (2008). Neuroprotective effects of prostaglandin A(1) in rat models of permanent focal cerebral ischemia are associated with nuclear factor-kappaB inhibition and peroxisome proliferator-activated receptor-gamma up-regulation, *J Neurosci Res.*, *86*, 1132-1141.

[113] Yan, X. D., Kumar, B., Nahreini, P., Hanson, A. J., Prasad, J. E. & Prasad, K. N. (2005). Prostaglandin-induced neurodegeneration is associated with increased levels of oxidative markers and reduced by a mixture of antioxidants, *J Neurosci Res.*, *81*, 85-90.

[114] Baek, J. Y., Jun, D. Y., Taub, D. & Kim, Y. H. (2003). Characterization of human phosphoserine aminotransferase involved in the phosphorylated pathway of L-serine biosynthesis, *Biochem J.*, *373*, 191-200.

[115] Diab, A., Deng, C., Smith, J. D., Hussain, R. Z., Phanavanh, B., Lovett-Racke, A. E., Drew, P. D. & Racke, M. K. (2002). Peroxisome proliferator-activated receptor-gamma agonist 15-deoxy-Delta(12,14)-prostaglandin J(2) ameliorates experimental autoimmune encephalomyelitis, *J Immunol.*, *168*, 2508-2515.

[116] Combs, C. K., Johnson, D. E., Karlo, J. C., Cannady, S. B. & Landreth, G. E. (2000). Inflammatory mechanisms in Alzheimer's disease: inhibition of beta-amyloid-stimulated proinflammatory responses and neurotoxicity by PPARgamma agonists, *J Neurosci.*, *20*, 558-567.

[117] Pierre, S. R., Lemmens, M. A. & Figueiredo-Pereira, M. E. (2009). Subchronic infusion of the product of inflammation prostaglandin J2 models sporadic Parkinson's disease in mice, *J Neuroinflammation.*, *6*, 18.

[118] Pena-dos-Santos, D. R., Severino, F. P., Pereira, S. A., Rodrigues, D. B., Cunha, F. Q., Vieira, S. M., Napimoga, M. H. & Clemente-Napimoga, J. T. (2009). Activation of peripheral kappa/delta opioid receptors mediates 15-deoxy-(Delta12,14)-prostaglandin J2 induced-antinociception in rat temporomandibular joint, *Neuroscience.*, *163*, 1211-1219.

[119] Cruz-Orengo, L., Dhaka, A., Heuermann, R. J., Young, T. J., Montana, M. C., Cavanaugh, E. J., Kim, D. & Story, G. M. (2008). Cutaneous nociception evoked by 15-deoxy-Delta12,14-PGJ2 via activation of ion channel TRPA1, *Mol Pain.*, *4*, 30.

[120] Takahashi, N., Mizuno, Y., Kozai, D., Yamamoto, S., Kiyonaka, S., Shibata, T., Uchida, K. & Mori, Y. (2008). Molecular characterization of TRPA1 channel activation by cysteine-reactive inflammatory mediators, *Channels*, (Austin)., *2*, 287-98.

In: Prostaglandins: Biochemistry, Functions… ISBN: 978-1-61668-272-9
Editors: Gillian M. Goodwin, pp.27-46

Chapter 2

PROSTAGLANDINS REGULATE GROWTH AND TRANSPORT IN CULTURED RENAL CELLS

Mary Taub*, E. James Springate and Facundo Cutuli
Department of Biochemistry and Department of Pediatrics,
School of Medicine and Biomedical Sciences, State University
of New York at Buffalo, Buffalo, New York, USA

1. ABSTRACT

Prostaglandins (PGs) are major regulators of growth and transport in kidney tubule epithelial cells. Although the kidney is a major producer of prostaglandins, little is understood about prostaglandin effects on tubule epithelial cells, and the signaling mechanisms involved in these effects. Our approach has been the use of in vitro kidney tubule epithelial cell culture systems in hormonally defined, serum free medium. We have found that Prostaglandin E_1 (PGE_1) and PGE_2 stimulate both growth and ion transport by an established kidney epithelial cell line, Madin Darby Canine Kidney, a distal tubule model. Similar affects on growth were observed with primary baby mouse kidney epithelial cells, indicating their physiologic significance. Subsequent to these studies, the overproduction of PGE_2 in the tumor microenvironment has been found to contribute to the formation of a number of carcinomas. This is because PGE_2 stimulates the growth of many types of epithelial cells in vivo. We have also found that PGE_1 and PGE_2 stimulate the Na,K-ATPase in MDCK cells, as well as primary rabbit kidney proximal tubule cells. Regulatory effects of PGE_1 and PGE_2 on Na,K-ATPase were attributed to transcriptional control. Prostaglandin Response Elements (PGREs) were identified on the Na,K-ATPase beta subunit promoter. These PGREs are highly conserved elements present on genomic DNA in a number of species. In addition, we have identified a novel regulatory mechanism for gene expression via these PGREs. Both cyclic AMP Regulatory Element Binding Protein (CREB) and Sp1 bind to PGREs, and interact so as to modulate the rate of transcription. The molecular mechanisms underlying the CREB/Sp1 interaction are being defined.

* Corresponding author: Biochemistry Dept, 140 Farber Hall, State University of New York at Buffalo, 3435 Main Street, Buffalo, NY 14214, Telephone: 716-829-3300, Fax: 716-829-2132, Email: biochtau@buffalo.edu

2. INTRODUCTION

Our laboratory has taken the in vitro approach to study the effects of PGs on the growth and function of kidney tubule epithelial cells. Our initial studies were conducted with an established canine kidney epithelial cell line, Madin Darby Canine Kidney (MDCK) [1]. In addition, we have used primary cultures to evaluate the physiologic significance of studies with MDCK [2] and to study properties of cells from other nephron segments [3].

MDCK cells are of interest to study in regards to their responsiveness to PGs, because of their close resemblance to normal distal tubule cells [1]. MDCK cells cultured on tissue culture polystyrene grow so as to form a confluent monolayer of polarized epithelial cells interconnected by tight junctions [4]. Domes (or hemicysts) can be visualized under an inverted microscope within confluent MDCK monolayers (Figure 1) [5]. When focusing on the entire monolayer, domes are groups of cells within the monolayer that are out of focus, being slightly elevated from the tissue culture dish. Dome formation is a consequence of the transepithelial transport of salt and water from the apical surface of the monolayer (which faces the tissue culture medium) through the basolateral surface (which faces the tissue culture plastic). The resulting hydrostatic pressure causes localized detachment of the monolayer from the tissue culture dish surface. When examining domes in MDCK monolayers by time-lapse photography, Leighton et al. [5] observed that the domes formed and burst, such that the monolayers appeared much like a bowl of bubbling oatmeal. Dome formation is a process that can be induced by a number of agents [6], including PGE_1 and PGE_2 [7] which are potent activators of Adenylate Cyclase (AC) in MDCK cells [8].

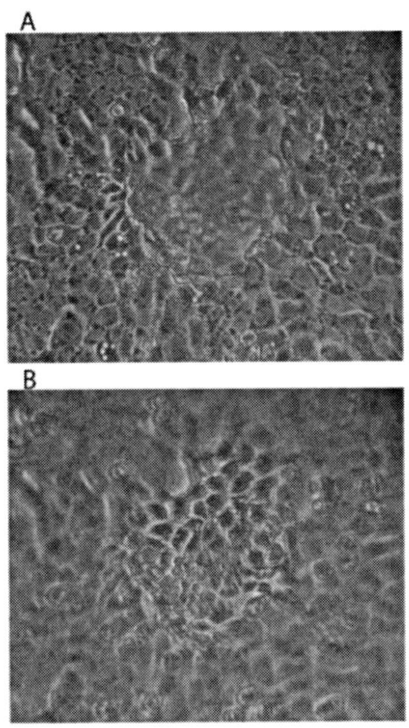

Figure 1. Dome Formation by MDCK Cells. Groups of MDCK cells slightly elevated from the culture dish when focusing on the monolayer (A), and when focusing of these same groups of cells (B)

In addition, primary cultures of kidney tubule epithelial cells are of interest to study. Having been directly removed from the animal, primary cultures closely resemble the cells in vivo both genotypically as well as phenotypically. Thus primary cultures can be used to examine the physiologic significance of studies with MDCK cells. In addition, primary cultures can be used to study the properties of cells derived from a number of nephron segments.

In this chapter we examine the regulation of the growth and transport functions of kidney tubule epithelial cell cultures by Prostaglandins (PGs). Hormonally defined medium is employed in these studies, because the effects of PGs can be examined reproducibly at levels that are present physiologically. In addition to studies with MDCK cells, studies are conducted with primary cultures of kidney tubule epithelial cells in defined medium. The use of serum free culture procedures permits primary cultures of kidney tubule epithelial cells to grow and retain their differentiated functions in the absence of fibroblast overgrowth.

3. Regulation of Kidney Epithelial Cell Growth by Prostaglandins

3A. Development of Defined Medium for MDCK Cells

Our first evidence concerning the important role of PGs in regulating MDCK cell growth was obtained while developing a hormonally defined serum free medium for this cell line [2]. Our results indicate that MDCK cells can grow in serum free medium supplemented with 5 μg/ml bovine insulin, 5 μg/ml human transferrin, 25 ng/ml PGE_1, 5 x 10^{-12} M triiodothyronine, and 5 x 10^{-8} M hydrocortisone (Medium K-1) at the same rate as in serum supplemented medium [2]. Moreover MDCK can grow continuously with passaging in serum free Medium K-1 for months. PGE_1 and PGE_2 were both growth stimulatory serum free, unlike the case with PGF_2 α. The growth stimulatory effects of PGE_1 and PGE_2 were optimal at a very low PGE concentration, 25 ng/ml (0.71 μM).

3B. Primary Culture Studies Provide a Rationale for Investigations with MDCK

Our interest in PGs arose initially from studies indicating dramatic effects of PGE_1 and PGE_2 on MDCK growth. Included amongst these studies were hormone deletion studies. In such a study, the effect of individually deleting each of the 5 supplements in Medium K-1 on growth was examined, and compared with the growth obtained with all the supplements (to account for synergistic effects between all of the supplements) [2]. The hormone deletion studies indicated that the individual deletion of PGE_1 from the medium had a more deleterious effect on growth, than the individual deletion of the other supplements. A similar result was obtained when a similar hormone deletion study was conducted with primary cultures of baby mouse kidney cells [9]. This latter result indicates that epithelial cells exist in the kidneys of normal mice that respond to hormones, growth factors and autocoids in a manner similar to MDCK. These observations were of particular interest, as the kidney is a major producer of PGs [10].

3C. Studies Conducted with Variants Altered in Their Responsiveness to PGs

At the time that the defined medium for MDCK cells was developed, most investigations concerning growth control were conducted with fibroblasts in serum-supplemented medium. Evidence had been obtained with cultured fibroblasts that increases in cAMP were growth inhibitory, rather than growth stimulatory. However these studies were conducted with high levels of cAMP analogs (and other agents which affect cAMP including PGs), which were required in order to observe effects of these agents in serum-supplemented medium. In contrast our studies in serum free medium revealed that low levels of a number of effector molecules including PGE_1, PGE_2 as well as vasopressin, glucagon, 8Br-cAMP and Isobutyl Methylxanthine (IBMX), a cAMP phosphodiesterase (PDE) inhibitor, were all growth stimulatory to MDCK cells. Thus, the presence of serum in the culture medium had masked the effects of agents that caused lower increases in cAMP (as components in serum already had increased intracellular cAMP to optimal levels for growth).

In order to study the effects of PGE_1 and PGE_2 further, we isolated Prostaglandin Independent (PGE_1 Ind) MDCK cells from mutagenized MDCK monolayers that no longer require either PGE_1 or PGE_2 for growth (normal MDCK cells eventually die in Medium K-1 lacking either PGE_1 or PGE_2). When a number of PGE_1 independent clones of MDCK were studied, PGE_1 and PGE_2 were found to be growth inhibitory at PGE concentrations as low as 25 ng/ml. PGE_1 independence can be explained by the increased intracellular cAMP level, and reduced cAMP PDE activity of the variant cells. In order to determine whether these alterations observed in vitro affected growth in vivo, the PGE_1 independent variants were injected into adult nude mice. Tumor formation was observed in the adult nude mice, unlike the case with normal MDCK cells. These results could be interpreted as indicating that the PG requirement for the growth of kidney epithelial cells was a limiting factor for normal growth in vivo.

In contrast to the results obtained with the PGE_1 independent variants, dibutyryl cAMP resistant (DB^r) MDCK variant cells retain the growth stimulatory effect of PGE_1 and PGE_2, although as discussed below transport is altered [7]. DB_r variant clones were isolated from mutagenized MDCK cells by their resistance to killing by 1 mM dibutyryl cAMP and 0.5 mM IBMX [7]. The DB_r variant MDCK cells have reduced Type I cAMP dependent protein kinase (A Kinase) activity [7]. These results can thus be explained if Type II A Kinase mediates the growth stimulatory effect of PGE_1 and PGE_2, while Type I A kinase mediates the effects of PGE_1 and PGE_2 on transport function (Figure 2). Type II A Kinase remains intact in DB^r cells, permitting them to grow in response to PGE_1. The Type II A Kinase activity of PGE_1 independent variants is activated in the absence of PGE_1, due to the increased cAMP levels of this latter class of variants.

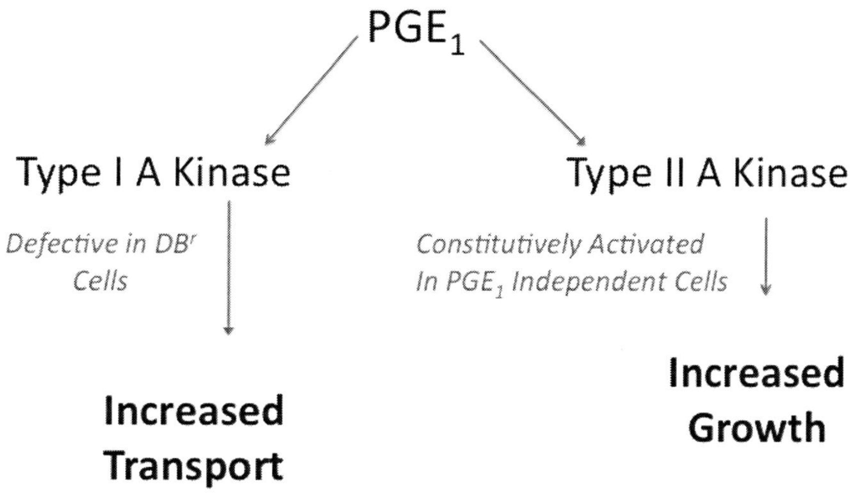

Figure 2. Effect of PGE_1 on Growth and Function is Mediated by A Kinase. PGE_1 stimulates both MDCK transport (Na,K-ATPase activity) and growth by means type I and type II A Kinase, respectively. A defect in type I A kinase in DB^r variants of MDCK cells results in a loss of the stimulatory effect of PGE_1 on transport, without affecting the stimulatory effect on growth. The increased intracellular cAMP in PGE_1 independent variants of MDCK cells results in a loss of the requirement for PGE_1 for optimal growth, without affecting the requirement for PGE_1 for increased transport by Na,K-ATPase.

3D. Regulation of Kidney Epithelial Cell Growth by Prostaglandins In Vivo

A number of in vivo studies support the contention that our investigations with PGs and MDCK cell growth are physiologically significance. Following a unilateral nephrectomy, hypertrophy and hyperplasia is observed in the remaining kidney. The involvement of PGs in this process is supported by the observations of Logan et al [11] that the compensatory hypertrophy in the remaining kidney is inhibited by indomethacin. Also consistent with the involvement of PGs, is the report of Pelayo and Shanley [12] that following nephrectomy, PG biosynthesis is increased in the remaining kidney. The increased PG biosynthesis has been attributed to the induction of COX-2 in the remaining kidney [13]. The underlying mechanism is not yet understood, although the COX-2 gene is included amongst the early response genes in cell cycle progression [14]

Moreover, the growth promoting effects of PGE_2 are not restricted to epithelial cells in the kidney. The serum concentration of PGE_2 is increased following partial hepatectomy [15]. Liver regeneration is impaired in animals treated with either COX1 or COX2 inhibitors [15]. PGE_1 and PGE_2 are growth stimulatory to primary rat hepatocytes in serum free medium [16]. The investigations of Meisdalen et al.[17] indicate that Phospholipase C (PLC) activation is involved in mediating the growth stimulatory effects of PGE_2 rather than AC. In addition evidence has been presented for the involvement of PGE_2 in regulating the growth of rat fetal lung epithelial cells [18], mouse intestinal crypt cells [19], and rat thyroid epithelial cells [20].

Recent studies have shown that the level of COX-2 is elevated in certain cancers including colorectal carcinoma, squamous cell carcinoma of the head and neck, as well as particular lung and breast cancers [21]. COX-2 levels have also been observed to be elevated in pre-cancerous lesions such as familial adenomatous polyposis that result in colorectal carcinoma. Thus, COX-2 and PGs have been proposed to play a role in carcinogenesis. PGE_2 in particular has been found to be growth stimulatory to colon cancer cells [22], while PGA_1 and PGJ_2 have been observed to oppose tumorigenesis [23]. The PGE_2-induced trophic affects on the gastrointestinal mucosa and on colon tumor cells involve transactivation of the EGF receptor [24]. After PGE_2 interacts with its specific receptors in HT-29 cells, the EGFR is rapidly phosphorylated and activated, resulting in activation of the ERK2 signaling pathway. Evidence has been presented that PGE_2 promotes angiogenesis in tumors, as well as metastasis [25].

4. Regulation of Kidney Epithelial Cell Function by PGs

4A. Summary of Functional Effects of Prostaglandins in the Kidney

PGs are produced in all regions of the kidney, the major prostanoid being PGE_2, followed by PGI_2, $PGF_2\alpha$ and TXA_2 [26]. Renal functions regulated by PGs include renin secretion, glomerular filtration, vascular tone and renal hemodynamics [27]. In addition, PGE_2 plays a key role in regulating transport by tubule epithelial cells. However, the effects of PGE_2 on other nephron segments are not well defined, and underlying mechanisms are not well understood. Thus we have employed the in vitro approach for mechanistic studies.

4B. Transepithelial Transport Studies with MDCK Cells

The MDCK cell line is a simplified model system amenable for a detailed molecular analysis of the signaling pathways responsible for eliciting observed effects of PGs in vitro. MDCK cells can be grown on semi-permeable supports to study their polarized morphology, and transport functions. The MDCK monolayers on permeable supports can be placed in Ussing chambers for electrophysiological studies [4, 28], and yield results similar to those obtained with frog skins or gallbladders. These studies indicate that MDCK cells have an electrical resistance across the monolayer (80-4000 Ω cm^2), which is associated with the conductance of a current across the monolayer by means of a transcellular as well as a paracellular pathway. The conductance across the transcellular pathway depends upon apical and basolateral transport systems, while the paracellular route depends upon occluding junctions. After MDCK cells are plated in tissue culture dishes, junction formation occurs. MDCK cells assume a polarized morphology, which is a prerequisite for trans-epithelial transport.

In order to characterize the individual Na^+ entry and efflux pathways involved in transepithelial Na^+ transport in MDCK cells, we conducted labeled solute uptake studies. Our uptake studies indicated that Na^+ enters MDCK cells via an amiloride-sensitive Na^+/H^+ antiport system, and leaves MDCK cells via a basolateral Na,K-ATPase, which is subject to inhibition by ouabain. Consistent with this hypothesis, the potential difference across MDCK monolayers is reduced by both amiloride and ouabain. Similarly, dome formation in confluent MDCK monolayers is inhibited by either 10^{-4} M amiloride or 0.5 μM ouabain [29]. In contrast, the frequency of domes in MDCK monolayers increases in the presence either PGE_1, 8BrcAMP, or IBMX, agents that increase intracellular cAMP levels. PGE_1, 8BrcAMP and IBMX may increase dome formation as a consequence of their effects on ion transport. However, PGE_1, 8BrcAMP and IBMX may also increase dome formation by affecting junctional complexes, as well as cell adhesion to the substratum.

4C. Regulation of Na,K-ATPase Activity by Prostaglandins

A major function of the mammalian kidney is to maintain the ionic composition and volume of the extracellular compartment within narrow limits. One manner in which the kidney performs this function, is through renal Na,K-ATPase. Na^+ in the lumen of the nephron is transported into tubule epithelial cells by apical ion transport systems. Subsequently, the basolateral Na,K-ATPase transports 3 intracellular Na^+ out of the cells by hydrolyzing ATP, while allowing 2 extracellular K^+, to enter the cells [30-33]. A consequence of transport by Na/K-ATPase, is that a sodium gradient is established across the plasma membrane, creating the driving force for other Na^+ dependent transport systems. Thus Na,K-ATPase acts as a "cornerstone of kidney transport machinery [34]." However, renal Na,K-ATPase is not simply a housekeeping enzyme, but instead is an important target of regulation, participating in the adaptations, which occur in response to environmental changes [34]. Hormonal regulation of Na,K-ATPase has been extensively studied [35]. In addition, a number of studies indicate that locally produced renal effector molecules also regulate Na,K-ATPase [36]. However, the effects of PGs on renal Na,K-ATPase are not well understood, given the complexity of the kidney.

Thus, the effects of PGE_1 and PGE_2 on the activity of the Na,K-ATPase were examined in MDCK cells. Following a 4 day incubation with either PGE_1 or 8BrcAMP, an increase in the initial rate of ouabain sensitive Rb^+ uptake was observed. The increased rate of ouabain sensitive Rb^+ uptake was associated with an increase in Na,K-ATPase activity, as well as an increase in the level of total cellular Na,K-ATPase (as determined by a Western analysis). In order to determine whether the increased level of Na,K-ATPase could be explained by an increased rate of biosynthesis, ^{35}S-methionine labeling studies were conducted. ^{35}S-Na,K-ATPase was immunoprecipitated (IPd), separated by SDS PAGE. The radioactive bands on the polyacrylamide gels which corresponded to the Na,K-ATPase α and β subunits were visualized by autoradiography. The autoradiographs indicated that the level of newly synthesized ^{35}S-Na,K-ATPase had increased in MDCK cells treated with either PGE_1 or 8BrcAMP. Thus, we concluded that PGE_1, PGE_2 and 8BrcAMP increased the rate of biosynthesis of Na,K-ATPase in MDCK cells.

The Na,K-ATPase consists of an α subunit, responsible for catalytic activity, as well as a β subunit, required for integration into the basolateral membrane. The α and β subunits of Na,K-ATPase are individually synthesized, followed by assembly in the Golgi apparatus. In MDCK cells, the level of the newly synthesized β subunit is limiting for heterodimer formation (while excess α is degraded) [37]. Thus, an increase in the level of newly synthesized Na,K-ATPase β subunit would be expected to cause a substantial increase in the level of Na,K-ATPase [38].

The results of our Northern analyses of α and β subunit mRNAs in MDCK cells are consistent with the importance of the β subunit as being a determinant of the cellular Na,K-ATPase level in MDCK cells. A stimulatory effect of PGE_1 on the Na,K-ATPase β subunit mRNA was observed at a PGE_1 concentration as low as 25 ng/ml (0.071 μM). The stimulatory effect of PGE_1 on the level of β subunit mRNA increased more than 10 fold, as the PGE_1 concentration increased to 500 ng/ml (1.4 μM) [39]. However the Na,K-ATPase α subunit mRNA level increase only 1.4 fold at the highest PGE_1 concentration studied (500 ng/ml). For this reason, in our subsequent studies concerning transcriptional regulation, the β subunit gene ATP1B1 was studied. Our initial transient transfection studies conducted with pHβ1-1141Luc, a human Na,K-ATPase promoter luciferase construct, indicated that the transcription was elevated 8 fold following a 4 hour incubation with either PGE_1 or PGE_2.

4D. Regulation in PGE_1 Independent and Dibutyryl cAMP Resistant MDCK Cells

Possibly, the stimulatory effect of PGE_1 and PGE_2 on Na,K-ATPase is simply an indirect effect resulting from the effects of PGE_1 and PGE_2 on cell growth and viability. However, the stimulatory effects of PGE_1 and PGE_2 on Na,K-ATPase were retained in PGE_1 independent MDCK cells which no longer require PGE_1 for long term growth [40]. This includes stimulatory effects of PGE_1 on the biosynthesis of Na,K-ATPase, and on transcription (by pH β1-1141Luc) [41].

The stimulatory effects of PGE_1 and PGE_2 on Na,K-ATPase activity were no longer observed in DB^r MDCK cells defective in Type I A Kinase [7]. The stimulatory effects of PGE_1 and PGE_2 on the biosynthesis of Na,K-ATPase were also lost in DB^r MDCK cells [41]. Thus, these results indicate that the effects of PGE_1 and PGE_2 on the biosynthesis of the Na,K-ATPase are dependent upon Type I A Kinase (Figure 3).

However, our transient transfection studies with pHβ1-1141 Luc indicate that the stimulatory effect of PGE_1 on ATP1B1 transcription is retained in DB^r cells [41]. These latter results indicate that transcriptional regulation by PGE_1 depends upon Type II A Kinase (which is intact in DB^r cells), in addition to regulation by PKC (as described below) (Figure 3). Presumably, the defect in Type I A Kinase in DB^r cells prevents the stimulatory effects of PGE_1 and PGE_2 on the biosynthesis of the Na,K-ATPase, by preventing yet another regulatory effect of PGE_1 which occurs at the post-transcriptional level (Figure 3). Na,K-ATPase β subunit mRNA is AU rich, and contains 3 AUUA motifs [42] recognized by AUF1 and HUR, which affects mRNA stability and/or the efficiency of translation of such mRNAs [43].

4E. In vivo Evidence of Prostaglandin Effects

When Na^+ levels increase in the body, the level of COX2 increases in the medullary interstitium, resulting in an increase in medullary PGE_2 production. The functional result is the inhibition of H_2O reabsorption in the medullary Collecting Duct (CD), as well as the inhibition of NaCl reabsorption in the Thick Ascending Limb (TAL) [44]. These effects of PGE_2 are prevented by treatment with indomethacin [45].

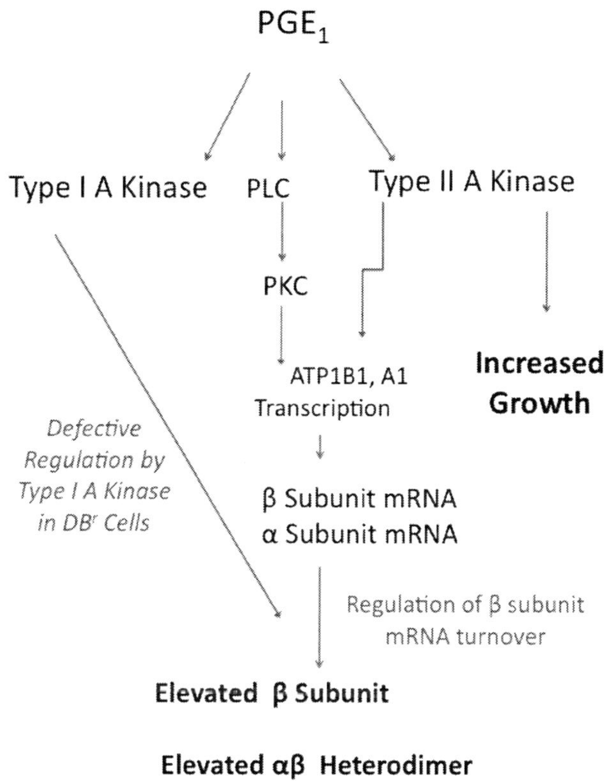

Figure 3. Effect of PGE_1 on Transport is Mediated by both Type I and Type II A Kinase. PGE_1 stimulates transcription of the ATP1B1 gene via type II A kinase (as observed in DBr cells which are defective in Type I A Kinase). However the defect in Type I A Kinase prevents the stimulatory effect of PGE_1 on the biosynthesis of the α and β subunits of Na,K-ATPase. This can be explained by a defect in regulation of post-transcriptional control which depends upon type I A kinase. The β subunit mRNA has AU sequences at its 3 prime end which are recognized by proteins involved in mRNA turnover and in the initiation of translation.

Another set of PG effects occur when dietary Na^+ is restricted. Under these conditions, the level of COX2 increases in the Macula Densa (MD) [46] [47] [48] [49]. The PGE_2 that is produced affects renal cells in the locale of the MD, a region spacially adjacent to the glomerulus. The specific cell types affected by these renal PGs are not always easy to ascertain. However, tubulopathies such as Bartter's Syndrome have given some indications. Normally, Na^+ reabsorption by the renal Na,K-2Cl^- cotransport system 2 (NKCC2) in the TAL occurs in excess of water. As a consequence, luminal fluid delivered to the Distal

Convoluted Tubule (DCT) is hypotonic. In patients with Bartter's Syndrome, the luminal fluid does not similarly become hypotonic. These patients have an adaptive response, increased production of PGE_2 by the MD. The PGE_2 (which is overproduced) stimulates distal tubule Na,K-ATPase, while inhibiting Na,K-ATPase in the CD, which contributes to the salt wasting which occurs in this disease [33] [48, 50, 51] [52]. Here it is also of interest to speculate that PGs may also contribute to the hypertrophy/hyperplasia of juxtaglomerular cells seen in Bartter's Syndrome.

4F. G Protein Coupled Receptors (GPCRs) That Mediate Effects of Prostaglandins

The effects of PGE_2 on transport can be attributed to signaling through EP receptors (GPCRs for PGE_1 and PGE_2) [53]. Four distinct classes of EP receptors are present in the kidney, including Gq-coupled EP1 (which activate PLC), Gs-coupled EP2 and EP4 (which activate AC), and Gi coupled EP3 (which inhibits AC) [54]. While all of the specific roles of the EP receptors have not yet been determined, their importance has been indicated by studies with transgenic animals. EP1 receptor knockout mice exhibit hypotension, while EP2 receptor knockout mice exhibit salt-sensitive hypertension, indicating their importance in Na^+ homeostasis [55].

In order to determine initially which classes of PG receptors regulate ATP1B1 gene expression, the effects of PGE_1, PGE_2 and $PGF_2\alpha$ were examined in transient transfection studies with pHβ1-1141 Luc. Both PGE_1 and PGE_2 (0.070-1.4 μM) were maximally stimulatory (8-fold) after 4 hr, unlike $PGF_2\alpha$, which had no effect. Our previous results indicated the involvement of cAMP in mediating the stimulatory effect of PGE_1 on Na,K-ATPase. These results indicate that a Gs-coupled EP receptor was involved. Consistent with the involvement of Gs-coupled EP2 are our observations that the EP2 selective agonist butaprost was stimulatory to transcription, while the EP2 selective antagonist AH6809 was inhibitory to the PGE_1 stimulation [56]. Also consistent with the involvement of Gs-coupled EP2, the PGE_1 stimulation is inhibited by the PKAI vector (encoding for the PKA inhibitory protein), as well as by PKI (a myristoylated PKA inhibitory peptide).

However, our results also indicate that a second type of EP receptor, Gq-coupled EP1, is involved in mediating the effects of PGE_1 and PGE_2 on transcription [56]. Consistent with the involvement of EP1 was stimulatory effect of the EP1 selective agonist 17-phenyl trinor PGE_2, as well as the inhibitory effect of the EP1 selective antagonist SC-51089 on the PGE_1 stimulation. Additional results consistent with the involvement of EP1 include the inhibition of the PGE_1 stimulation by the PKCI vector (encoding for the inhibitory domain of PKC) as well as by the PKCα and PKCβ1 specific inhibitors Bisindolylmaleimide I and Gö6976 [56, 57]. Phorbol 12-myristate 13-acetate (TPA) and 8BrcAMP were also stimulatory.

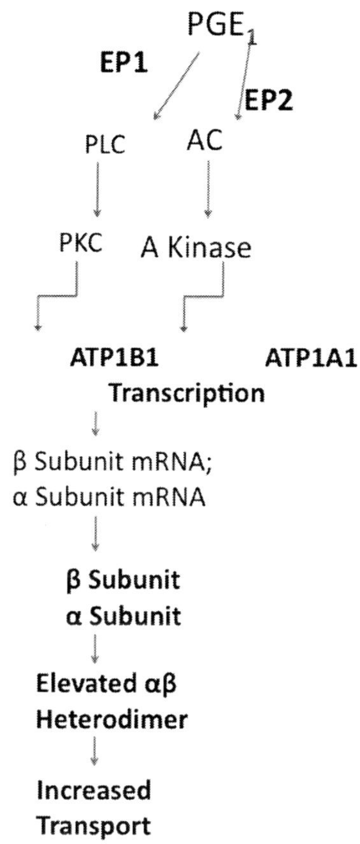

Figure 4. Effect of PGE_1 on Transcription is Mediated by EP1 and EP2 Receptors. Both Gq coupled EP1 receptors and Gs coupled EP2 receptors mediate the effect of PGE_1 on ATP1B1 transcription. The activation of both PKC (by EP1 mediated signaling) is necessary for increased transcription of the human ATP1B1 gene. The consequence is an increase in the level of the β subunit mRNA and to some extent even the α subunit mRNA at higher PGE_1 concentrations. The consequence is an increase in the rate of biosynthesis of Na,K-ATPase, resulting in an increase in the cellular level of Na,K-ATPase, as well as the level in the basolateral membrane. The ultimate result is an increase in the activity of the Na,K-ATPase in MDCK cells.

Thus, to summarize these results, PGE_1 stimulates transcription by activating EP1 receptors (resulting in PLC and ultimately PKC activation), as well as EP2 receptors (resulting in the activation of AC and A Kinase) (Figure 4). The activation of type I A kinase is responsible for mediating the post-transcriptional affects of PGE_1 on Na,K-ATPase (as stated above). In contrast, the activation of type II A kinase is responsible for mediating the stimulatory affects of PGE_1 on the transcription of the Na,K-ATPase ATP1B1 gene (Figure 4). In addition, as stated above, the growth stimulatory effect of PGE_1 is mediated by Type II A kinase.

4G. Transcriptional Regulation of Na,K-ATPase by Prostaglandins

Two distinct, differentially regulated genes, ATP1A1 and ATP1B1 encode for the Na,K-ATPase α1 and β1 subunits in the kidney [58].These 2 genes, ATP1A1 and ATP1B1 [59], are highly conserved, including the promoters which share >80% identity cross-species, suggestive of common modes of regulation, and validating in vitro animal model systems derived from dog, rat, mouse and rabbit. Included amongst the regulatory elements present on both the ATP1A1 and ATP1B1 promoters are Mineralocorticoid/Glucocorticoid Regulatory Elements (MREs/GREs), and Thyroid hormone Regulatory Elements (TREs). The structure of the human ATP1B1 promoter is shown in Figure 5.

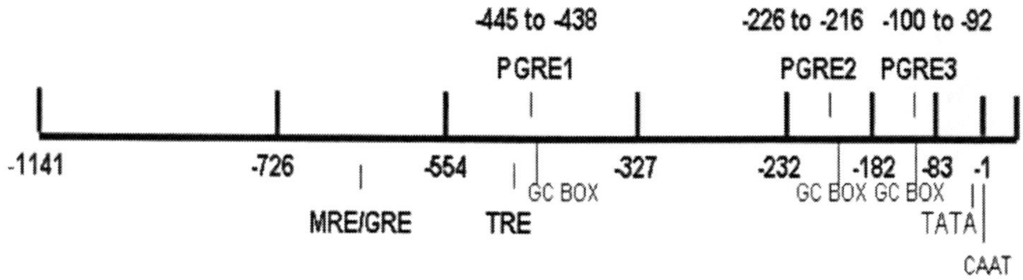

Figure 5. Regulatory Elements in the Human ATP1B1 Promoter. Regulatory elements on the ATP1B1 promoter include a mineralocorticoid, glucocorticoid regulatory element (MRE/GRE), a Thyroid Hormone Regulatory Element (TRE), Sp1 sites (i.e. GC boxes), a CAAT box, a TATA box and Prostaglandin Regulatory Elements (PGREs). The location of 5' deletion mutations also Indicated underneath. The stimulatory effect of PGE_1 was retained in pHβ1- 182Luc, unlike pHβ1-82Luc consistent with a requirement for PGRE3.

4G1. Identification of Prostaglandin Regulatory Elements

Initially, we postulated that the PGE stimulation depends upon 3 regulatory elements within the ATP1B1 promoter, PGRE1 (AGTCCCTGC, -445 to -438), PGRE2 (GTCCCTCA -226 to -216), and PGRE3 (AGTCCCTGC, -100 to -92). The requirement of these response elements in order to elicit the PGE_1 response was initially evaluated by transient transfection studies with 5' promoter deletion mutants (Figure 5). In addition studies were conducted with mutants with internal deletions. The stimulatory effect of PGE_1 8Br-cAMP, TPA were retained in deletion constructs which still contain PGRE3, unlike deletion mutant pHâ1-83 Luc, lacking PGRE3. When the 5' deletion mutations removed the MRE/GRE, the TRE, PGRE1 and PGRE2, the PGE stimulation was retained, provided that the 5' deletion construct contained PGRE3. However, the internal deletion Δ 72-299 (which removed both PGRE2 and PGRE3) (Figure 6) was also stimulatory, suggesting that PGRE1 also functions as a prostaglandin regulatory.

To examine this further, PGRE ODNs were inserted into the minimal promoter of pLuc-MCS, and the effect of PGE_1 examined. The results of these studies indicated that a PGE_1 stimulation of more than 80 fold could be obtained with a pLucMCS vector containing either PGRE1 or PGRE3 [60, 61]. However the stimulatory effect of PGE_1 was no longer observed when the sequence of PGRE3 was altered. A similar result was obtained in the case of PGRE1.

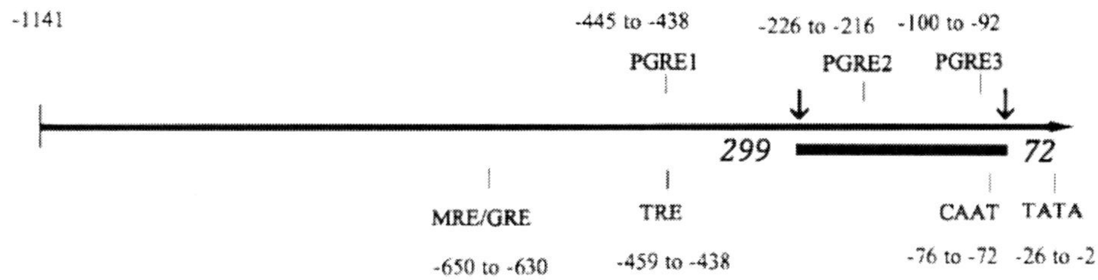

Figure 6. Effect if an internal deletion mutation (-299 to -72) in the ATP1B1 promoter. The stimulatory effect of PGE_1 was retained in the deletion mutant pHβ1-1141Luc Δ-299 to -72, which retained PGRE1 but lacked PGRE2 and PGRE3.

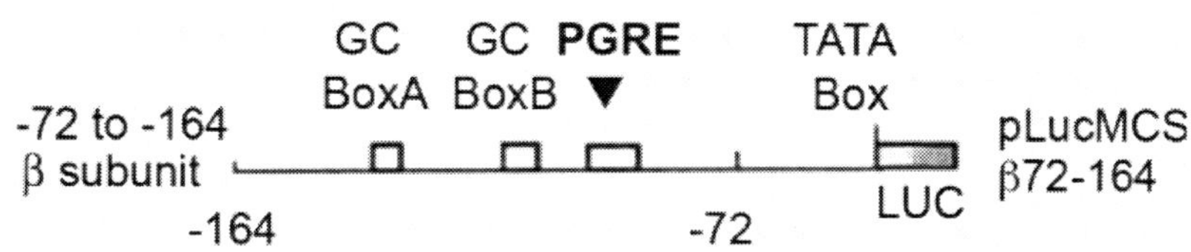

Figure 7. The PGRE3 Region in Human ATP1B1. pLucβ72-164 contains a region of the human ATP1B1 promoter (-72 to -163) with 2 GC boxes, PGRE3, and a TATA box [1].

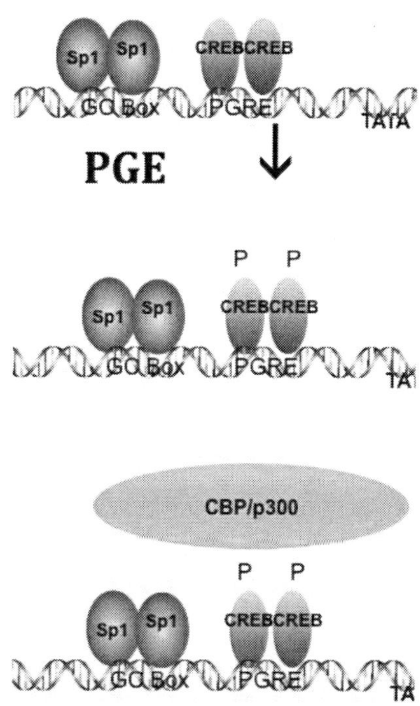

Figure 8. Model of Transcriptional Regulation by the ATP1B1 Gene. PGE_1 causes the phosphorylation of CREB bound to PGRE3 on the human ATP1B1 promoter. As a consequence CBP becomes associated with CREB in the promoter region. Adjacent Sp1 facilitates the binding of CBP to CREB, the formation of the pre-initiation complex, and the initiation of transcription of the ATP1B1 gene.

The role of the Sp1 site adjacent to PGRE3 was examined using the vector pLucMCSβ 72-164 (Figure 7). When the *2 Sp1 Sites* (GC boxes) immediately upstream of PGRE3 (Figure 7) were translocated further upstream from PGRE3 (pLucMCS β72-164 2xGC trans), the stimulatory effect of PGE_1 was lost. Similarly, mutations in the Sp1 site (CCCGCCC→ AATTACA) resulted in a loss of the PGE_1 stimulation. The involvement of GC boxes was studied further using pLucMS-FN, which contains a 4 GC boxes within the human fibronectin (FN) promoter (-167 to -72). When PGRE3 was inserted in between these 4 GC boxes, a large PGE_1 stimulation was obtained, indicating that PGRE3 is functional in such a heterologous promoter.

4G2. Model for Transcriptional Regulation by Trans-acting Factors

We have proposed a model to explain the mechanism by which PGE_1 and PGE_2 stimulate transcription of the ATP1B1 gene (Figure 8). cAMP Regulatory Element (CRE) Binding Protein (CREB), a homodimer, binds to the PGRE3 site, in close proximity to Sp1, a homodimer, which binds to an adjacent upstream GC box. After MDCK cells are treated with PGE_1, CREB, which is bound to PGRE sites, is phosphorylated, and activated. The CREB Binding Protein (CBP) then interacts with phosphorylated CREB (pCREB), so as to facilitate the initiation of transcription. The CREB/CBP interaction is very likely stabilized by Sp1 (either via an Sp1/CREB interaction, or an Sp1/CBP interaction).

The results of our Electrophoretic Mobility Shift Assays (EMSAs) and DNA affinity precipitation studies are consistent with this hypothesis. In the EMSAs, ^{32}P-labeled PGRE ODNs were used in binding reactions with nuclear extracts from MDCK cells. After ^{32}P-labeled PGRE/nuclear protein complexes formed, the ^{32}P-ODN/protein complexes were subjected polyacrylamide gel electrophoresis and the bands visualized by autoradiography. The migration of ^{32}P-ODN/protein complexes was retarded as compared with free, unbound ^{32}P-ODN. When an excess of either an unlabeled consensus CRE oligo (TGACGTCA), or an unlabeled consensus Sp1 oligo (CCGCCC) was added to the binding reaction, the intensity of the ^{32}P-labeled bands decreased, suggesting that the unlabeled oligos compete with ^{32}P-labeled oligo for binding to either CREB or Sp1, respectively. When either an anti-Sp1 antibody or an anti-CREB antibody was also added to the binding reaction, the migration was retarded even further, and supershifts were observed. These latter observations are also consistent the binding of both CREB and Sp1 to the ^{32}P-PGRE3 oligo. In the DNA affinity precipitation studies, biotinylated PGRE oligos were similarly incubated with nuclear extracts from MDCK cells, purified by Streptavidin sepharose precipitation, and subjected to a Western analysis. The Western analysis indicated that CREB, CBP and Sp1 all co-precipitate with the same biotinylated PGRE3 ODN. Also consistent with this hypothesis, GST-CREM (a homolog of CREB) pulled down Sp1 and CBP from lysates of PGE1-treated MDCK cells. CREB and CBP were pulled down with Sp1 IPs, and similarly Sp1 and CBP were pulled down with CREB IPs.

CREB binds to CREs present in a number of genes in the mammalian genome. After mammalian cells are treated with agents that activate PKA and/or PKC, CREB becomes phosphorylated while bound to these CREs, which often causes CBP/p300 to be recruited, and as a consequence results in CREB activation. However, a recent genome-wide analysis conducted by Zhang et al. [62], indicates that CREB activation cannot always be attributed simply to CREB phosphorylation and the recruitment of CBP. These investigators proposed that additional CREB binding partners are involved in activating CREB. Our studies with the ATP1B1 gene in particular indicate that Sp1 acts as such a CREB binding partner, in addition to CBP/p300.

To our knowledge such CREB/Sp1 interactions are unique characteristics of the ATP1B1 and ATP1A1 promoters. However, several other genes, including Chromogranin A, have adjacent CRE and Sp1 sites [63] [64]. Recently, evidence has been presented for the involvement of PGE_2 in Chromogranin A gene regulation [65]. PGE_2 has also been shown to be involved in regulation of transcription of Human T-Cell Leukemia Virus Type 1 (HTLV-1) long terminal repeats (LTRs) [66]. Transcriptional regulation of HTLV-1 LTRs depends upon an interaction between Sp1, CREB and Tax, a specific HTLV-1 coactivator [67] [68, 69], and in this respect, has similarities with ATP1B1 gene regulation.

4H. Prostaglandin Effects on Renal Proximal Tubule Function

The effects of PGE effects that we observed on transcription in MDCK cells may be indicative of a regulatory effect of PGE_1 specific to the renal distal tubule. Alternatively, tubule epithelial cells derived from other nephron segments may show evidence of a similar responsiveness to PGs. In order to examine this hypothesis, similar transient transfection studies were conducted with primary renal proximal tubule cell cultures. Our laboratory has previously developed a primary rabbit kidney proximal tubule cell culture system in serum free medium that retains many of the differentiated functions and hormone responses typical of renal proximal tubule cells. Our results indicate that PGE_1 an PGE_2 similarly stimulate transcription of Na,K-ATPase in primary RPT cells. However our results indicate that Gs-coupled EP4 receptors are involved, unlike the case with MDCK cells.

The results of our studies with primary RPT cells are of particular interest. For some time it was thought that PGs did not play a major role in regulation of renal proximal tubule function. However, recently this view has changed [27]. PGE_2 inhibits Phosphate (Pi) transport in mouse proximal convoluted tubules [70]. In vitro studies with the opossum kidney cell line OK, indicate that PGE_2 synthesized by RPT cells acts via EP4 receptors so as to stimulate the activity of the basolateral p-aminohippurate (pAH) transport system in the RPT, so as to increase the secretion of organic anions into the lumen of the nephron [71]. The physiologic significance is indicated by studies with isolated rabbit S2 proximal tubules, in which similar results were obtained. Interestingly, PGE_2 is a substrate of the pAH transport system. PGE_2 secretion is inhibited following a long-term exposure of RPTs to PGE_2, due to a chronic inhibitory effect of PGE_2 on this transport system.

5. Conclusion

PGs affect both the growth and functional properties of kidney tubule epithelial cells. The in vitro approach has been taken to determine the signaling mechanisms involved in mediating these effects. PGE_1 and PGE_2 both stimulate growth, as well as Na,K-ATPase activity of the established MDCK cell line, a distal tubule model system. The involvement of cAMP in mediating the growth stimulatory effects of PGE_1 is indicated studies with PGE1 independent variant cells. The involvement of Type I A kinase in mediating post-transcriptional effects of PGE1 on Na,K-ATPase is indicated by studies with dibutyryl cAMP resistant MDCK cells (defective in Type I A kinase). Transcriptional regulation of the Na,K-ATPase beta subunit gene (ATP1B1) involves EP1 as well as EP2 receptors. Transcriptional regulation of ATP1B1 involves the activation of PKC as well as Type II A kinase (which is intact in the DBr MDCK cells). Two prostaglandin response elements (PGREs) have been identified in the ATP1B1 promoter. CREB binds to PGREs. Regulation by PGE_1 and PGE_2 involves the phosphorylation of CREB, the recruitment of CBP, as well as an interaction of Sp1 with the CREB/CBP complex. Our studies with primary RPT cells indicate that this mode of regulation also occurs in the proximal tubule, and thus is not unique to the distal tubule. In the proximal tubule, Gs coupled EP4 receptors are involved in mediating the effects of PGE_1 and PGE_2, in addition to Gs-coupled EP2 and Gq coupled EP1 receptors.

REFERENCES

[1] Saier, MH; Jr., Boerner, P; Grenier, FC; McRoberts, JA; Rindler, MJ; Taub, M; et al. Sodium entry pathways in renal epithelial cell lines. *Miner Electrolyte Metab.*, 1986, 12(1), 42-50.

[2] Taub, M; Chuman, L; Saier, MH; Jr., Sato, G. Growth of Madin-Darby canine kidney epithelial cell (MDCK) line in hormone-supplemented, serum-free medium. *Proc Natl Acad Sci*, U S A. 1979, Jul, 76(7), 3338-42.

[3] Chung, SD; Alavi, N; Livingston, D; Hiller, S; Taub, M. Characterization of primary rabbit kidney cultures that express proximal tubule functions in a hormonally defined medium. *J Cell Biol.*, 1982 Oct;95(1), 118-26.

[4] Cereijido, M; Robbins, ES; Dolan, WJ; Rotunno, CA; Sabatini, DD. Polarized monolayers formed by epithelial cells on a permeable and translucent support. *J Cell Biol.*, 1978, Jun;77(3), 853-80.

[5] Leighton J; Brada Z; Estes LW; Justh G. Secretory activity and oncogenicity of a cell line (MDCK) derived from canine kidney. *Science.* 1969 Jan 31, 163(866), 472-3.

[6] Lever JE. Inducers of mammalian cell differentiation stimulate dome formation in a differentiated kidney epithelial cell line (MDCK). Proc *Natl Acad Sci U S A.* 1979 Mar, 76(3), 1323-7.

[7] Devis PE; Grohol SH; Taub M. Dibutyryl cyclic AMP resistant MDCK cells in serum free medium have reduced cyclic AMP dependent protein kinase activity and a diminished effect of PGE_1 on differentiated function. *J Cell Physiol.* 1985 Oct;125(1), 23-35.

[8] Rindler MJ; Chuman LM; Shaffer L; Saier MH; Jr. Retention of differentiated properties in an established dog kidney epithelial cell line (MDCK). *J Cell Biol.* 1979 Jun;81(3), 635-48.

[9] Taub M; Sato G. Growth of functional primary cultures of kidney epithelial cells in defined medium. *J Cell Physiol.* 1980 Nov;105(2), 369-78.

[10] Morel F. Sites of hormone action in the mammalian nephron. *Am J Physiol.* 1981 Mar; 240(3), F159-64.

[11] Logan JL; Benson B. Serum renotropic factor stimulates prostaglandin synthesis in

primary cultures of rabbit kidney cells. *Prostaglandins Leukot Essent Fatty Acids*. 1990 Nov;41(3), 183-6.

[12] Pelayo JC; Shanley PF. Glomerular and tubular adaptive responses to acute nephron loss in the rat. Effect of prostaglandin synthesis inhibition. *J Clin Invest*. 1990 Jun;85(6), 1761-9.

[13] Wang JL; Cheng HF; Zhang MZ; McKanna JA; Harris RC. Selective increase of cyclooxygenase-2 expression in a model of renal ablation. *Am J Physiol*. 1998 Oct;275(4 Pt 2), F613-22.

[14] Fletcher BS; Kujubu DA; Perrin DM; Herschman HR. Structure of the mitogen-inducible TIS10 gene and demonstration that the TIS10-encoded protein is a functional prostaglandin G/H synthase. *J Biol Chem*. 1992 Mar 5;267(7), 4338-44.

[15] Rudnick DA; Perlmutter DH; Muglia LJ. Prostaglandins are required for CREB activation and cellular proliferation during liver regeneration. *Proc Natl Acad Sci U S A*. 2001 Jul 17;98(15), 8885-90.

[16] Kimura M; Osumi S; Ogihara M. Prostaglandin E(2) (EP(1)) receptor agonist-induced DNA synthesis and proliferation in primary cultures of adult rat hepatocytes: the involvement of TGF-alpha. *Endocrinology*. 2001 Oct;142(10), 4428-40.

[17] Meisdalen K; Dajani OF; Christoffersen T; Sandnes D. Prostaglandins enhance epidermal growth factor-induced DNA synthesis in hepatocytes by stimulation of E prostanoid 3 and F prostanoid receptors. *J Pharmacol Exp Ther*. 2007 Sep;322(3), 1044-50.

[18] Jassal D; Han RN; Caniggia I; Post M; Tanswell AK. Growth of distal fetal rat lung epithelial cells in a defined serum-free medium. *In Vitro Cell Dev Biol*. 1991 Aug;27A(8), 625-32.

[19] Cohn SM; Schloemann S; Tessner T; Seibert K; Stenson WF. Crypt stem cell survival in the mouse intestinal epithelium is regulated by prostaglandins synthesized through cyclooxygenase-1. *J Clin Invest*. 1997 Mar 15;99(6), 1367-79.

[20] Di Paola R; Menzaghi C; De Filippis V; Corda D; Di Cerbo A. Cyclooxygenase-dependent thyroid cell proliferation induced by immunoglobulins from patients with Graves' disease. *J Clin Endocrinol* Metab. 1997 Feb;82(2), 670-3.

[21] Rishikesh MK; Sadhana SS. Prostaglandins and Cyclooxygenase: Their Probable Role in Cancer. *Indian Journal of Pharmacology*. [Review]. 2003;35, 3-12.

[22] Fischer SM. Prostaglandins and cancer. *Front Biosci*. 1997; 2, d482-500.

[23] Ishikawa T; Akimaru K; Nakanishi M; Tomokiyo K; Furuta K; Suzuki M; et al. Anti-cancer-prostaglandin-induced cell-cycle arrest and its modulation by an inhibitor of the ATP-dependent glutathione S-conjugate export pump (GS-X pump). *Biochem J*. 1998 Dec 15;336 (Pt 3), 569-76.

[24] Pai R; Soreghan B; Szabo IL; Pavelka M; Baatar D; Tarnawski AS. Prostaglandin E_2 transactivates EGF receptor: a novel mechanism for promoting colon cancer growth and gastrointestinal hypertrophy. *Nat Med*. 2002 Mar; 8(3), 289-93.

[25] Harris RE. Cyclooxygenase-2 (cox-2) and the inflammogenesis of cancer. *Subcell Biochem*. 2007;42, 93-126.

[26] Qi Z; Cai H; Morrow JD; Breyer MD. Differentiation of cyclooxygenase 1- and 2-derived prostanoids in mouse kidney and aorta. *Hypertension*. 2006 Aug;48(2), 323-8.

[27] Nasrallah R; Clark J; Hebert RL. Prostaglandins in the kidney, developments since Y2K. *Clin Sci* (Lond). 2007 Oct;113(7), 297-311.

[28] Misfeldt DS; Hamamoto ST; Pitelka DR. Transepithelial transport in cell culture. *Proc Natl Acad Sci U S A*. 1976 Apr;73(4), 1212-6.

[29] Rindler MJ; Taub M; Saier MH; Jr. Uptake of $_{22}Na^{+}$ by cultured dog kidney cells (MDCK). *J Biol Chem*. 1979 Nov 25;254(22), 11431-9.

[30] Bonvalet JP; Pradelles P; Farman N. Segmental synthesis and actions of prostaglandins along the nephron. *American Journal of Physiology*. 1987;253(3 Pt 2), F377-87.

[31] Stokes JB; Kokko JP. Inhibition of sodium transport by prostaglandin E_2 across the isolated, perfused rabbit collecting tubule. *Journal of Clinical Investigation*. 1977;59(6), 1099-104.

[32] Culpepper RM; Andreoli TE. PGE_2, forskolin, and cholera toxin interactions in modulating NaCl transport in mouse mTALH. American *Journal of Physiology*. 1984;247(5 Pt 2), F784-92.

[33] Scherzer P; Wald H; Rubinger D; Popovtzer MM. Indomethacin and sodium retention in the rat, role of inhibition of prostaglandin E_2 synthesis. *Clinical Science*. 1992;83(3), 307-11.

[34] Feraille E; Doucet A. Sodium-potassium-adenosinetriphosphatase-dependent sodium transport in the kidney: hormonal control. *Physiological Reviews*. 2001;81(1), 345-418.

[35] Lingrel JB; Orlowski J; Shull MM; Price EM, Department of Molecular Genetics B, Microbiology UoCCoMO. *Molecular genetics of Na,K-ATPase.* Progress in nucleic acid research and molecular biology. 1990;38.

[36] Aperia AC. Intrarenal dopamine: a key signal in the interactive regulation of sodium metabolism. *Annual review of physiology*. 2000;62, 621-47.

[37] Geering K. The functional role of the beta-subunit in the maturation and intracellular transport of Na,K-ATPase. FEBS Lett. 1991 Jul 22;285(2), 189-93.

[38] kathi g. The functional role of the beta-subunit in the maturation and intracellular transport of sodium-potassium ATPase. *FEBS.* 1991; 285(2), 189-93.

[39] Taub M; Borsick M; Geisel J; Rajkhowa T; Allen C. Regulation of the Na,K-ATPase in MDCK cells by prostaglandin E^1: a role for calcium as well as cAMP. *Experimental Cell Research.* 2004; 299, 1-14.

[40] Taub ML; Wang Y; Yang IS; Fiorella P; Lee SM. Regulation of the Na,K-ATPase activity of Madin-Darby canine kidney cells in defined medium by prostaglandin E^1 and 8-bromocyclic AMP. *J Cell Physiol.* 1992 May;151(2), 337-46.

[41] Borsick M; Rajkhowa T; Taub M. Evidence for post-transcriptional regulation of Na,K-ATPase by prostaglandin E^1. *Biochem Biophys Res Commun.* 2006 Jun 30;345(2), 739-45.

[42] Shao Y; Ismail-Beigi F. Control of Na^{+}-K^{+}-ATPase beta 1-subunit expression: role of 3'-untranslated region. *American Journal of Physiology - Cell Physiology*. 2004;286(3), C580-5.

[43] Giles KM; Daly JM; Beveridge DJ; Thomson AM; Voon DC; Furneaux HM; et al. The 3'-untranslated region of p21WAF1 mRNA is a composite cis-acting sequence bound by RNA-binding proteins from breast cancer cells, including HuR and poly(C)-binding protein. *The Journal of biological chemistry*. 2003 Jan 31;278(5), 2937-46.

[44] Stokes JB. Integrated actions of renal medullary prostaglandins in the control of water excretion. *Am J Physiol.* 1981 Jun;240(6), F471-80.

[45] Dunn MJ; Hood VL. Prostaglandins and the kidney. American Journal of Physiology. 1977;233(3), 169-84.

[46] Paliege A; Mizel D; Medina C; Pasumarthy A; Huang YG; Bachmann S; et al. Inhibition of nNOS expression in the macula densa by COX-2-derived prostaglandin E(2). *Am J Physiol Renal Physiol.* 2004 Jul;287(1), F152-9.

[47] Harris RC; Breyer MD; George MOBK; Urologic Diseases C, Division of Nephrology DoMVUSoM, Veterans Affairs Medical Center NTUSARHmve. Physiological regulation of cyclooxygenase-2 in the kidney. *American journal of physiology Renal physiology.* 2001 Jul;281(1), F1-11.

[48] Jeck N; Schlingmann KP; Reinalter SC; Komhoff M; Peters M; Waldegger S; et al. Salt handling in the distal nephron: lessons learned from inherited human disorders. *American Journal of Physiology - Regulatory Integrative & Comparative Physiology.* 2005;288(4), R782-95.

[49] Amirlak I; Dawson KP. Bartter syndrome: an overview. *Qjm.* 2000;93(4), 207-15.

[50] Pedemonte CH; Efendiev R; Bertorello AM. Inhibition of Na,K-ATPase by dopamine in proximal tubule epithelial cells. *Semin Nephrol.* 2005 Sep;25(5), 322-7.

[51] Tang MJ; Wang YK; Lin HH. Butyrate and TGF-beta downregulate Na,K-ATPase expression in cultured proximal tubule cells. *Biochem Biophys Res Commun.* 1995 Oct 4;215(1), 57-66.

[52] Magyar CE; Zhang Y; Holstein-Rathlou NH; McDonough AA. Downstream shift in sodium pump activity along the nephron during acute hypertension. *J Am Soc Nephrol.* 2001 Nov;12(11), 2231-40.

[53] Matlhagela K; Taub M. Involvement of EP1 and EP2 receptors in the regulation of the Na,K-ATPase by prostaglandins in MDCK cells. *Prostaglandins & Other Lipid Mediators.* 2006;79(1-2), 101-13.

[54] Breyer MD; Breyer RM. G protein-coupled prostanoid receptors and the kidney. *Annu Rev Physiol.* 2001;63, 579-605.

[55] Breyer MD; Breyer RM. Prostaglandin E receptors and the kidney. *Am J Physiol Renal Physiol.* 2000 Jul;279(1), F12-23.

[56] Matlhagela K; Taub M. Involvement of EP1 and EP2 receptors in the regulation of the Na,K-ATPase by prostaglandins in MDCK cells. *Prostaglandins & other lipid mediators.* 2006 Mar;79(1-2), 101-13.

[57] Taub M; Borsick M; Geisel J; Matlhagela K; Rajkhowa T; Allen C. Regulation of the Na,K-ATPase in MDCK cells by prostaglandin E^1: a role for calcium as well as cAMP. *Exp Cell Res.* 2004 Sep 10;299(1), 1-14.

[58] Lingrel JB; Orlowski J; Shull MM; Price EM. Molecular genetics of Na,K-ATPase. *Progress in Nucleic Acid Research & Molecular Biology.* 1990;38, 37-89.

[59] Lingrel JB; Kuntzweiler T. Na^+,$K(^+)$-ATPase. *J Biol Chem.* 1994 Aug 5;269(31), 19659-62.

[60] Matlhagela K; Taub M. Regulation of the Na-K-ATPase beta(1)-subunit promoter by multiple prostaglandin-responsive elements. *Am J Physiol Renal Physiol.* 2006 Sep;291(3), F635-46.

[61] Matlhagela K; Borsick M; Rajkhowa T; Taub M. Identification of a prostaglandin-responsive element in the Na,K-ATPase beta 1 promoter that is regulated by cAMP and Ca^{2+}. Evidence for an interactive role of cAMP regulatory element-binding protein and Sp1. *Journal of Biological Chemistry.* 2005;280(1), 334-46.

[62] Zhang X; Odom DT; Koo SH; Conkright MD; Canettieri G; Best J, et al. Genome-wide analysis of cAMP-response element binding protein occupancy, phosphorylation, and

target gene activation in human tissues. *Proc Natl Acad Sci U S A.* 2005 Mar 22;102(12), 4459-64.

[63] Shell SA; Fix C; Olejniczak D; Gram-Humphrey N; Walker WH. Regulation of cyclic adenosine 3',5'-monophosphate response element binding protein (CREB) expression by Sp1 in the mammalian testis. *Biol Reprod.* 2002 Mar;66(3), 659-66.

[64] Hèocker M; Raychowdhury R; Plath T; Wu H; O'Connor DT, Wiedenmann B, et al. Sp1 and CREB mediate gastrin-dependent regulation of chromogranin A promoter activity in gastric carcinoma cells. *J Biol Chem.* 1998 Dec 18;273(51), 34000-7.

[65] Connolly R; Gates D; Loh N; Baban D; Thakker R; Johnston B; et al. Cox-2 promotes chromogranin A expression and bioactivity: evidence for a prostaglandin E_2-dependent mechanism and the involvement of a proximal cyclic adenosine 5'-monophosphate-responsive element. *Endocrinology.* 2007 Sep;148(9), 4310-7.

[66] Moriuchi M; Inoue H; Moriuchi H. Reciprocal interactions between human T-lymphotropic virus type 1 and prostaglandins: implications for viral transmission. *J Virol.* 2001 Jan;75(1), 192-8.

[67] Mayr BM; Guzman E; Montminy M. Glutamine rich and basic region/leucine zipper (bZIP) domains stabilize cAMP-response element-binding protein (CREB) binding to chromatin. *J Biol Chem.* 2005 Apr 15;280(15), 15103-10.

[68] Asahara H; Santoso B; Guzman E; Du K; Cole PA; Davidson I; et al. Chromatin-dependent cooperativity between constitutive and inducible activation domains in CREB. *Mol Cell Biol.* 2001 Dec;21(23), 7892-900.

[69] Suzuki T; Kimura A; Nagai R; Horikoshi M. Regulation of interaction of the acetyltransferase region of p300 and the DNA-binding domain of Sp1 on and through DNA binding. *Genes Cells.* 2000 Jan;5(1), 29-41.

[70] Syal A; Schiavi S; Chakravarty S; Dwarakanath V; Quigley R; Baum M. Fibroblast growth factor-23 increases mouse PGE_2 production in vivo and in vitro. *Am J Physiol Renal Physiol.* 2006 Feb;290(2), F450-5.

[71] Sauvant C; Holzinger H; Gekle M. Modulation of the basolateral and apical step of transepithelial organic anion secretion in proximal tubular opossum kidney cells. Acute effects of epidermal growth factor and mitogen-activated protein kinase. *Journal of Biological Chemistry.* 2001; 276(18), 14695-703.

In: Prostaglandins: Biochemistry, Functions… ISBN: 978-1-61668-272-9
Editors: Gillian M. Goodwin, pp.47-63

Chapter 3

COMPARATIVE EFFECTS OF PGE_2 ON DIFFERENT TYPES OF IMMUNE CELLS IN HOLOTHURIAN *EUPENTACTA FRAUDATRIX*

L. S. Dolmatova*[*] *and O. A. Zaika
Il'ichev Pacific Oceanological Institute, Far Eastern Branch, Russian Academy of Sciences, Baltiyskaya Street 43, Vladivostok, 690041 Russia

ABSTRACT

Tissues of certain holothurians (Echinodermata, Holothuroidea, Dendrochirota) are known to contain relatively high levels of prostaglandins (PGs) including PGE_2. In contrast to vertebrates, the contributions of PGs to immunoregulation in echinoderms have been little explored. In this study, we investigated the specifics in effects of PGE_2 on apoptosis in different types of immune cells in holothurian *Eupentacta fraudatrix*. The studies were conducted in different cell fractions obtained after density gradient centrifugation: the fraction of phagocytes of higher density (P1), the fraction of phagocytes of lower density (P2), the fraction combined of two phagocyte types (CP), and fraction enriched in morula cells (MC). DNA fragmentation was detected by agarose gel electrophoresis and by Hoechst 33342 staining. All fractions studied demonstrated pronounced apoptotic background with specific intensity for each of them. Inhibitor of endogenous PG synthesis indomethacin (10 μg/g) administration one day prior to experiments inhibited apoptosis compared to its background value both in PC and MC. While indomethacin was used both in vivo (10 μg/g) and in vitro (10^{-6} M), the level of apoptosis insignificantly increased in freshly isolated MC, but not in PC. In 24 h incubation, PGE_2 (10^{-6} M) counteracted effects of indomethacin on apoptosis in MC but not in PC. In 48 h, both indomethacin (10^{-6} M) and the more combined indomethacin (10^{-6} M) plus PGE_2 (10^{-7}-10^{-6} M) decreased apoptosis both in MC and PC of animals pretreated with indomethacin in vivo. Dexamethasone (10^{-4}-10^{-7} M) induced apoptosis in the reverse concentration-dependent manner in PC in 24 h, and 10^{-4} M dexamethasone even inhibited apoptosis compared to the control. However, in P1, dexamethasone (10^{-4} M) significantly increased apoptosis compared to the control in 24 h. In 48 h,

[*] Corresponding author: E-mail: dolmatova@poi.dvo.ru

dexamethasone produced similar effects in P1 and MC, and the reverse dexamethasone-concentration dependence of apoptosis level was shown in both types of the cells. Commercially available catalase benefited apoptosis-modulating effects of both indomethacin and dexamethasone on apoptosis in fractions of phagocytes, but not in MC. Both indomethacin or dexamethasone (10^{-4} M) counteracted apoptosis-increasing effect of PG in normal MC, but failed to effect apoptosis reduction following PGE_2 (10^{-6} M) treatment in phagocytes. Apparently, PGE_2 may be involved in apoptosis modulation both in MC and phagocytes of holothurian *E. fraudatrix,* and different sensitivities of MC and phagocyte fractions to apoptosis-modulating effect of PG may play an important role in cooperation of immunocytes during immune response similar to that in vertebrates.

INTRODUCTION

Studying mechanisms of mammalian immunity and searching for the relatively simple models resulted in great attention from scientists around the world being directed to invertebrate immunity in recent years.

Being Deuterostomata, Echinoderms have a common ancestor with vertebrates [1], and therefore are the objects of special interest. The free cells of coelomic fluid (coelomocytes) of these animals are heterogenous and vary in their content in different species. Phagocytes are the most studied among other coelomocytes, and they are considered to be analogues of macrophages of vertebrates [2]. Morula cells, similar to phagocytes, are the most numerous and present among coelomocytes of all Echinoderms species. It is known that they participate in foreign material encapsulation and are involved in regeneration of connective tissue, transport and storage of nutrients [2], and are able to synthesize lectins [3], which apparently play a role of antibodies in invertebrates. However, functions of different types of coelomocytes are not studied completely, and the relationship between different types of coelomocytes during immune response remains confusing.

Apoptosis is one of the most important elements of immune response. Apoptosis—programmed cell death—allows the organism to eliminate differentiating or functionally deficient cells and cells potentially dangerous (mutant, infected, etc.) [4]. Studies of apoptosis in echinoderms demonstrated that it has many common features with that in vertebrates, e.g., it is accompanied by morphological changes of cells, chromatin condensation and degradation, and activation of caspases [5]. However, apoptosis and its control mechanisms remain poorly studied in immunocytes of Echinoderms.

Eicosanoids [postaglandins (PG) and leukotrines] are a group of oxygenated fatty acid derivatives formed from C20 polyunsaturated fatty acids, including arachidonic and eicosapentaenoic acids [6]. The key enzyme responsible for PG biosynthesis from arachidonic acid is cyclooxygenase (COX). In immune cells of vertebrates, PGE_2 plays the critical role in the regulation of immune response [7]. In particular, PGE_2 modulates apoptosis in immune cells via the induction of adenylate cyclase synthesis and cAMP increase [8] [9]. The studies on invertebrates were largely conducted on insects [10, 11], and the data obtained indicated the significance of eicosanoids in the immunity control, including apoptosis. It is notable, that prostaglandin-like compounds have been identified in many marine invertebrates [12,13]. The classical prostaglandins such as PGE_2 and $PGF_{2\alpha}$ were determined using radioimmunoassays in mollusks and in such echinoderms as sea urchins [14]. However, results of HPLC showed, that most of the immunologically active compounds found were not classical metabolites of arachidonic acid. On the other hand, classical prostaglandins, including PGE_2, were found later using a wide spectrum of modern methods of analysis, including HPLC, in many species of mollusks, corals, barnacles [6]. Thus, the nature of PG-like compounds of marine invertebrate tissues is under discussion.

The significance of PGs in immunity control of marine invertebrates also remains underexplored. The inflammation induction by PGE_1 was demonstrated in marine arthropods [15], and this also suggests that PGs can be involved in immunity regulation in marine invertebrates.

Among echinoderms, particularly high levels of PGs including PGE_2 were observed in holothurians (Holothuroidea, Dendrochirota) *Apostichopus japonicus* and *Cucumaria japonica*, and much less PGs were found in the tissues of related holothurian species *Eupentacta fraudatrix* [13]. At the same time, nearly no publications concerning the effect of PGs on echinoderm immunity and possible mechanisms of this effect are available.

Previously, two fractions of coelomocytes of *E. fraudatrix* enriched with phagocytes: up to 97 % (fraction 1, F1) and about 90% (fraction 2, F2) and morula cell-enriched fraction (MC) were obtained using a ficoll-verografin – step gradient centrifugation [16].

In mammals, phagocytes of two types-monocytes and neutrophils-differ in their PG-synthesizing activities: monocytes are the most source of PGs, and neutrophils mostly synthesize another product of metabolism of arachidonic acid-via the lipoxigenase way-leukotrienes [6]. Leukotrienes and prostaglandins, in turn, often play different roles in immunity regulation. The role of PGs in regulation of functional activity of different immunocytes of holothurians is not recognized. The aim of the present work was to study the effects of PGE_2 on the apoptosis in phagocytes and morula cells of holothurian *Eupentacta fraudatrix*.

Chemicals

Bovine serum albumin (BSA), EDTA, PGE_2, indomethacin, trypan blue, guanidine hydrochloride, ethidium bromide, Hoeschst 33342 were obtained from Sigma (St. Louis, Mo., USA). Ficoll 400 was obtained from Pharmacia (Sweden). Dexamethasone was purchased from KRKA (Slovenia). Other reagents were manufactured in Russia.

Animals

Holothurians *E. fraudatrix* (4-6 cm in length) were collected in Peter the Great Bay in autumn and winter 2002-2003. Prior to experiments, the animals were kept in an aquarium with running aerated seawater for not less than two weeks.

Coelomocyte Preparation

Holothurians were dissected, and coelomic fluid (1 V) was transferred into the bottles containing 2 V of anticoagulant solution (30 mmol /l EDTA, 31 g/l NaCl, 50 mmol/l Tris-HCl at pH 7.6). The coelomic fluid of 15-25 animals was combined for the subsequent cell fractionation.

Separation of Coelomocytes

Fluid samples were layered on the top of a step gradient, made up by mixing ficoll-verografin and anticoagulant solution in ratios of 1:2 (step 4), 1:1 (step 3), 1:0.4 (step 2), and 1:0 (step 1), v/v (modification of the method of Bertheussen & Seljelid [17]). The centrifugation was carried out at 300 g for 15 min at 5°C. After that, distinct bands at the interfaces were collected. The phagocyte fractions were collected from the coelomic fluid sample/step 4 (fraction P1) and step 4/step 3 (fraction P2) interfaces, and then they were combined (fraction PC) or in several cases remained separate. The fraction enriched in morula cells (MC) was collected from the step 2/step 3 interface. The cells were washed with buffered saline (PBS, 36 g/l, pH 7.4) and resuspended in medium 199 supplemented with 16.41 g/l NaCl, 0.264 g/l KCl, 0.87 g/l $CaCl_2$, 4.98 g/l $MgCl_2$ x $6H_2O$, 3.87 g/l $MgSO_4$ x $7H_2O$, 22.74 g/l glycine, 0.1 g/l glucose, 2.5 g/l BSA, and 50 mg/l sodium oxacillin [18].

The cell viability was quantified by trypan blue exclusion [19]. Cells were counted in a Goryaev chamber.

Experimental Protocol

In the first series of experiments, one day prior to in vitro experiments, endogenous PG biosynthesis was inhibited by indomethacin (10 μg/g) in PBS administration to animals. Control animals were injected with PBS. Several animals were remained intact.

The cell suspension of the isolated immunocyte fractions were incubated in round-bottom plates (5 x 10^5-1 x 10^6 cells/well) at 20°C with PBS (control), indomethacin (10^{-6} M) (control for indomethacin effect in vitro), or indomethacin (10^{-6} M) plus PGE_2 (10^{-6} , 10^{-7}, and 10^{-8} M). The incubation was performed in two parallel series. The samples were taken after 0, 24, and 48 h of incubation.

In another series of experiments, no in vivo administration of indomethacin was undertook. Aliquots (1 ml) of cell suspensions (1×10^6 cells / ml) were added into round-bottomed plates and incubated at 20°C with PGE_2 (10^{-6} M) or different concentrations of dexamethasone (0, 10^{-8} M, 10^{-7} M, 10^{-6} M, 10^{-4} M) or combination of PGE_2 (10^{-6} M) and indomethacin (10^{-6} M) or combination of PGE_2 (10^{-6} M) and dexamethasone (10^{-4} M). The samples of cell suspensions were collected after 0, 24, and 48 h of incubation (three sets of experiments).

In both series of experiments, commercially available catalase, mark B ("Reachim", Russia) from bovine (0.7 μg/ml) was added to the cells in several cases.

After finishing an experiment, a portion of cell suspension was centrifuged at 1000xg for 5 min at 5°C and the pellet was frozen in liquid nitrogen and stored at -18°C until DNA isolation. A portion of the cells (0.3 x 10^6 – 0.5 x 10^6) after centrifugation was fixed with 4% formalin.

Apoptosis Measuring

Apotosis was estimated by DNA fragmentation using DNA electrophoresis in agarose gel or by nuclear morphology changes by using Hoeschst 33342 staining.

Hoeschst 33342 Staining

Smears of samples fixed with formalin were made and dried at the air with further staining with Hoeschst 33342 as described by Komatsu et al. [20]. The percentage of apoptotic cells was defined using fluorescent microscope “Polyvar”(Germany).

DNA Isolation and Electrophoresis

Cell pellets were lysed in a buffer containing 5mM Tris-HCl, 0.5% Triton X-100 (w/v), 1 mM EDTA for 20 min at a room temperature. DNA deproteinization was performed with 4M guanidine hydrochloride in 0.05 M Tris-HCl containing 0.01 M EDTA [21]. DNA was treated with chloroform and isopropanol and precipitated with ethanol. Precipitate, after centrifugation at 10,000 rpm, was dissolved in TE buffer (Tris-HCl/ EDTA, pH 8.0). DNA samples were separated by electrophoresis in a 1% (w/v) agarose horizontal gel containing ethidium bromide.

Statistical Analysis

The results are expressed as the mean ± SE. The data were statistically analyzed using Student’s t-test. A P value<0.05 was taken as significant.

Results

The level of background apoptosis in normal fractions of immunocytes of holothurian *E. fraudatrix* was shown to decrease as follows: MC >P2> P1 (Figure 1).

Indomethacin (10 µg/g) injected to animals a day before an experiment abolished apoptosis in freshly isolated MC. In vivo, indomethacin diminished apoptosis also in freshly isolated PC (Figure 2). However, when indomethacin (10^{-6} M) was added to isolated cell fractions of animals, pretreated with indomethacin in vivo(10 µg/g), the level of apoptosis in MC, but not in PC, insignificantly increased just after addition compared to that in animals injected with indomethacin only in vivo.

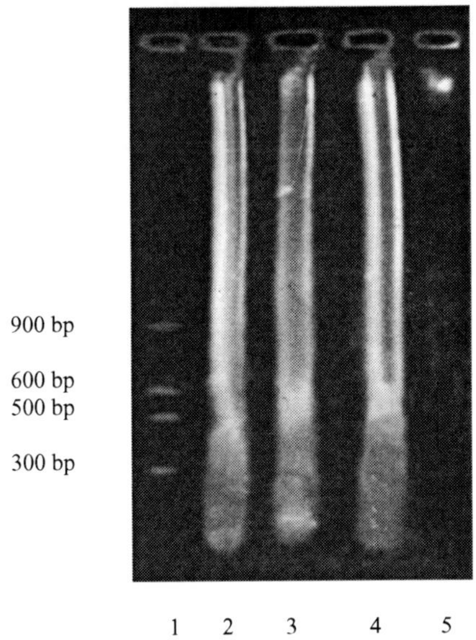

Figure 1. Apoptosis in freshly isolated fractions of coelomocytes of holothurians *Eupentacta fraudatrix* injected with buffered saline (lines 2-4) or indomethacin (10 μg/g, once a day before the experiment). Lines: 1-markers, 2-MC, 3 – P1, 4-P2, 5- MC treated with indomethacin.

Figure 2. Apoptosis in freshly isolated MC (lines 2, 4) and PC (lines 3, 5) of holothurians treated with indomethacin in vivo (10 μg/g, once a day) (lines 2 and 3) or in vivo and in vitro (10^{-6} M) (lines 4,5). Line 1-markers.

Nevertheless, pretreatment with indomethacin in vivo produced a reverse effect on apoptosis in MC if the cells were incubated after their isolation for 24-48 h. Thus, indomethacin (10 μg/g) slightly stimulated apoptosis (Figure 3A) and decreased cell viability by 36% in MC (Figure 3B) compared to the control (saline) in 24 h. PGE$_2$ (10^{-6} M), in contrast, decreased apoptosis compared to the control. Addition of commercially available catalase together with PGE$_2$ did not influence effect of PG.

A

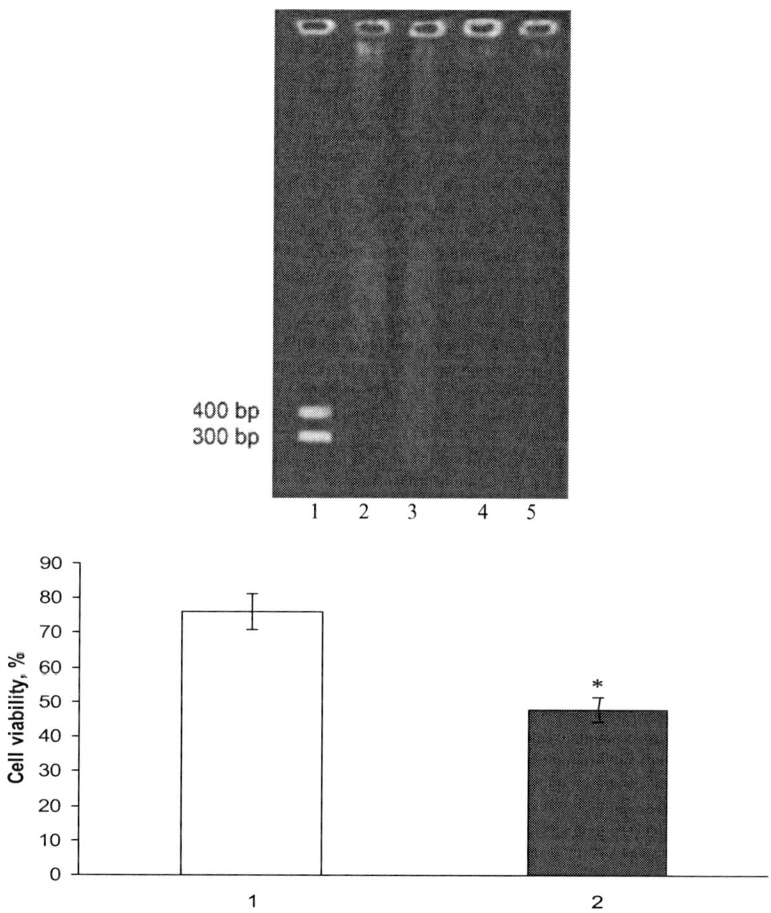

B

A lines: 1-markers, 2- control (injection with saline), 3- injection with indomethacin (10 μg/g, once a day), 4- injection with indomethacin + PGE$_2$ (10^{-6} M), 5- injection with indomethacin + PGE$_2$ (10^{-6} M) + catalase (0.7 μg/ml).

B: 1-control, 2- indomethacin (10 μg/g, once a day).

Figure 3. Effects of indomethacin, PGE$_2$, and commercially available catalase on the apoptosis (A) and cell viability after pretreatment with indomethacin (B) of MC of holothurian *Eupentacta fraudatrix* in 24 h of incubation.

The measuring of cell viability in PC showed that indomethacin (10 μg/g) decreased cell viability by 25% in 24 h-incubation, and PGE$_2$ (10^{-6} M) decreased cell viability by 55% (Figure 4). Commercially available catalase reduced effect of PGE$_2$.

The further 48 h-incubation of indomethacin (10^{-6} M) with MC (Figure 5A) of the animals pretreated with indomethacin (10 μg/g) also was not resulted in significant apoptosis. Combined administration of indomethacin (10^{-6} M) and PGE_2 (10^{-7}-10^{-6} M) further decreased the level of apoptosis, but increased it for 10^{-8} M PGE_2 compared to indomethacin alone.

In PC, level of apoptosis following treatment with indomethacin (10^{-6} M) and the more so with 10^{-7}-10^{-6} M PGE_2 was suppressed (Figure 5B) after 48 h-incubation.

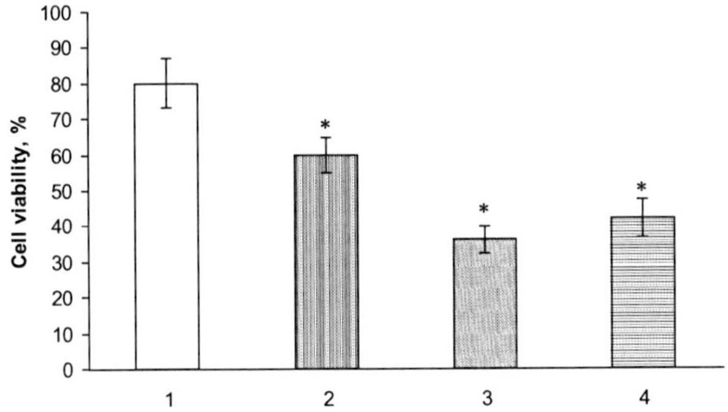

1-control (saline), 2- indomethacin (10 μg/g, once a day), 3- indomethacin (10 μg/g)+ PGE_2 (10^{-6} M), 4- indomethacin (10 μg/g)+PGE_2 (10-6 M)+ catalase (0.7 μg/ml).

* $P<0.05$ compared to control.

Figure 4. Effects of PGE2 and catalase on the cell viability of PC in holothurians pretreated with indomethacin in 24 h.

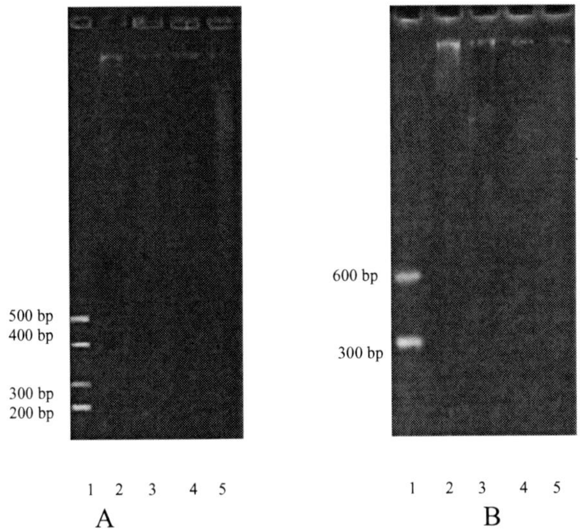

A lines: 1- markers, 2- indomethacin (10^{-6} M), 3- indomethacin (10^{-6} M) + PGE_2 (10^{-6} M), 4- indomethacin (10^{-6} M) + PGE_2 (10^{-7} M), 5- indomethacin (10^{-6} M) + PGE_2 (10^{-8} M).

B lines: 1- markers, 2-buffered saline, 3- indomethacin (10^{-6} M), 4- indomethacin (10^{-6} M) + PGE_2 (10^{-6} M), 5- indomethacin (10^{-6} M) + PGE_2 (10^{-7} M).

Figure 5. The influence of indomethacin (10^{-6} M) and different concentrations of PGE_2 on apoptosis in fractions of coelomoctyes (A-MC, B-PC) of holothurian *Eupentacta fraudatrix* pretreated with indomethacin in vivo (10 μg/g, once a day) after 48 h-incubation.

The changes in the apoptosis level corresponded to the changes in cell viability. For instance, after 48 h - incubation MC with 10^{-6} M PGE_2 together with indomethacin (10^{-6} M), cell viability was higher by 35% compared to indomethacin alone (Figure 6). On the contrary, using 10^{-8} M PGE_2 together with indomethacin decreased the viability of MC by 60% compared to indomethacin alone.

When another inhibitor of PG biosynthesis dexamethasone (10^{-4}-10^{-7} M) was added to normal PC, the reverse concentration-dependent changes in apoptosis were found in 24 h (Figure 7A), and dexamethasone (10^{-4} M) even inhibited apoptosis compared to the control. However, in pure fraction of phagocytes P1, dexamethasone (10^{-4} M) (Figure 7B), in contrast, significantly increased apoptosis compared to the control in 24 h. This process was accompanied with significant decrease in cell viability (Figure 7C).

Hoechst 33342 staining revealed that dexamethasone also induced apoptosis in reverse concentration-dependent manner in PC in 48 h (Figure 8). Dexamethasone (10^{-4} M) inhibited apoptosis, and the effect was more pronounced when dexamethasone was added together with catalase.

In 48 h, dexamethasone (10^{-4} M) insignificantly stimulated also apoptosis in MC compared to the control (Figure 9), and PGE_2 (10^{-6} M) slightly decreased apoptosis compared to that in dexamethasone-treated cells, and this effect of PG was maintained when PG was used together with dexamethasone (10^{-4} M). Indomethacin (10^{-6} M) administered together with PGE_2 (10^{-6} M) abolished apoptosis in this time interval.

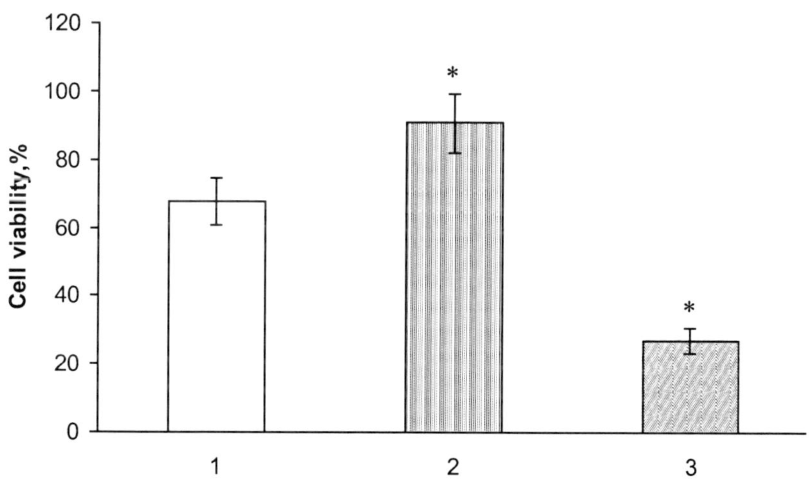

1- indomethacin (10^{-6} M), 2- indomethacin (10^{-6} M) + PGE_2 (10^{-6} M), 3- indomethacin (10^{-6} M) + PGE_2 (10^{-8} M).

* $P<0.05$ compared to indomethacin effect.

Figure 6. Effects of indomethacin (10^{-6} M) and PGE_2 on cell viability in MC of holothurians pretreated with indomethacin (10μg/g once a day) in 48 h.

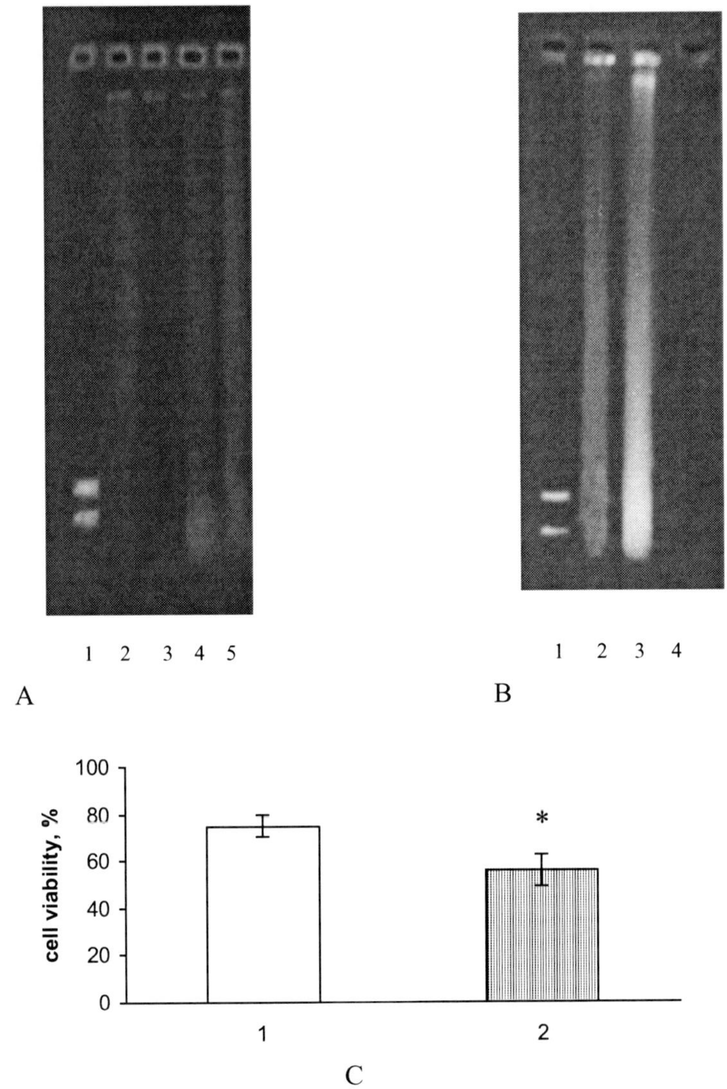

A lines: 1-markers (300 and 400 bp), 2-control (buffered saline), 3- dexamethasone (10^{-4} M), 4-dexamethasone (10^{-6} M), 5- dexamethasone (10^{-7} M), 6- dexamethasone (10^{-4} M) + catalase (0.7 μg/ml).

B lines: 1-markers (300 and 400 bp), 2-control (buffered saline), 3-dexamethasone (10^{-4} M), 4-dexamethasone (10^{-4} M) + catalase (0.7 μg/ml).

C: 1-control (buffered saline, 2- dexamethasone (10^{-4} M).

* $P<0.05$ compared to control.

Figure 7. Effects of dexamethasone and commercially available catalase on apoptosis in PC (A) and P1 (B) and cell viability in P1 (C) after 24 h-incubation.

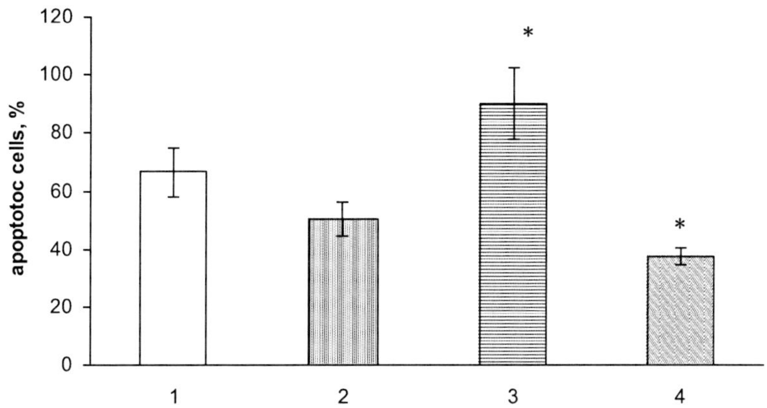

1-control, 2- dexamethasone (10^{-4} M), 3- dexamethasone (10^{-6} M), 4- dexamethasone (10^{-4} M) + catalase.

*-$P<0.05$ compared to the control.

Figure 8. Effects of dexamethasone and commercially available catalase (0.7 μg/ml) on apoptosis (Hoechst 33342 staining) in PC in 48 h.

1 2 3 4 5

Lines: 1-control, 2- dexamethasone (10^{-4} M), 3- PGE$_2$ (10^{-6} M), 4-dexamethasone (10^{-4} M) + PGE$_2$ (10^{-6} M), 5-indomethacin (10^{-6} M) + PGE$_2$ (10^{-6} M).

Figure 9. Effects of inhibitors of cyclooxygenase and PGE$_2$ on apoptosis of MC of normal holothurians in 48 h.

Data on a cell viability (Figure 10) indicate that dexamethasone (10^{-4}-10^{-6} M) did not effect the viability of MC in 48 h, and only the smallest concentration of dexamethasone (10^{-7} M) significantly decreased the viability of the cells (by 27%). This fact indicate that dexamethasone induces decrease in MC viability in a reverse concentration-dependent manner. Catalase alone or dexamethasone (10^{-4} M) plus catalase did not change MPC viability compared to the control, the same was true for PGE_2 (10^{-6} M) or PGE_2 (10^{-6} M) combined with indomethacin (10^{-6} M). However, when PGE_2 was used together with dexamethasone (10^{-4} M), cell viability significantly decreased (by 26%).

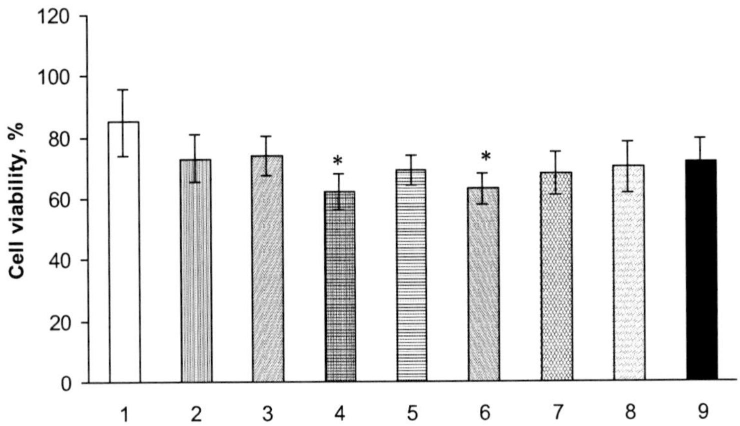

1-control, 2- dexamethasone (10^{-4} M), 3- dexamethasone (10^{-6} M), 4- dexamethasone (10^{-7} M), 5- dexamethasone (10^{-4} M) + catalase, 6- dexamethasone (10^{-4} M) + PGE_2 (10^{-6} M), 7- PGE_2 (10^{-6} M), 8-indomethacin + PGE_2 (10^{-6} M), 9- catalase.

*-$P<0.05$ compared to the control.

Figure 10. Influence of dexamethasone (10^{-4} M), PGE_2 (10^{-6} M), and commercially available catalase (0.7 μg/ml) on cell viability in MC in 48 h.

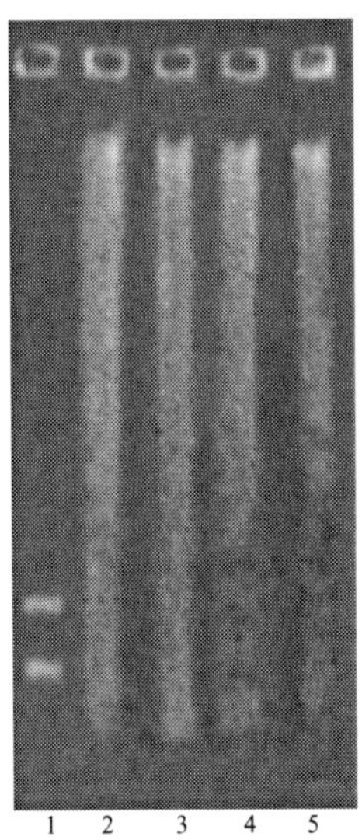

Lines: 1-markers (400 bp, 300 bp), 2- control, 3- PGE_2 (10^{-6} M), 4- dexamethasone (10^{-4} M), 5- dexamethasone (10^{-4} M) + PGE_2 (10^{-6} M).

Figure 11. Effects of PGE_2 and dexamethasone on apoptosis in normal P2 after 48 h-incubation.

In normal P2, dexamethasone (10^{-4} M), but not PGE_2 (10^{-6} M), increased apoptosis level compared to that in control in 48 h (Figure 11), but dexamethasone (10^{-4} M) and PGE_2(10^{-6} M), being added to the cells together, induced lower level of apoptosis compared to that induced by dexamethasone alone.

CONCLUSION

The level of apoptosis in freshly isolated cells of marine invertebrates (spontaneous or background apoptosis) varies between species and cell types [22], and its level can be considerable in some echinoderms [5, 23]. In this chapter, we demonstrated that the level of spontaneous apoptosis in freshly isolated fractions of holothurian coelomocytes is also high and exceeds the level commonly observed in leukocytes of vertebrates. The coelomocyte fractions studied differed in their apoptosis levels: MC >P2> >P1. This may indicate the presence of different functional activities of those fractions, and/or different regulatory mechanisms. PGs, and PGE_2 especially, play an important role in regulation and cooperation of immune response in vertebrates [7], and separate types of leukocytes differ in their ability to synthesize PGs. To test the influence of PGs on the apoptosis level in immunocytes of holothurian *E. fraudatrix*, we firstly used the inhibitor of PG biosynthesis indomethacin.

In spite of the existing controversy on the nature and mechanism of biosynthesis of PGs in marine invertebrates [6], it is known from a number of reports that PG biosynthesis in wide range of invertebrates, including Deuterostomata, is inhibited by indomethacin [24], suggesting the presence of COX-derived mechanism of PG biosynthesis in animals studied. We showed that indomethacin injected in vivo (10 μg/g, once a day) decreased level of apoptosis both in freshly isolated CP and MC compared to the control.

Apoptosis-decreasing effect of indomethacin administered in vivo was changed to the apoptosis-increasing one after 24 h incubation of MC. Changes in apoptosis level corresponded to the changes in cell viability of MC. PGE_2 (10^{-6} M) effect on apoptosis in MC pretreated with indomethacin in vivo counteracted indomethacin influence. However, data on cell viability in PC in 24 h indicate that the effects of PGE_2 (10^{-6} M) and indomethacin (10 μg/g) were unidirectional, and both of them, apparently, stimulated apoptosis in this fraction in this time interval. The data obtained indicate that apoptosis- modulating effect of indomethacin is time-dependent. In addition, it seems clear that MC and PC vary in their sensitivity to exogenous PGE_2. Thus, PGE_2 effects MC in a direct manner, which is inhibited by indomethacin, and PC-in reverse manner, in accordance with feedback mechanism, and indomethacin does not inhibit this process. The further support to these suggestions comes from the data on addition of indomethacin (10^{-6} M) to the cells pretreated with the preparation in vivo. While indomethacin was used both in vivo (10 μg/g) and in vitro (10^{-6} M), the level of apoptosis insignificantly increased in freshly isolated MC, but not in PC.

In 48 h-incubation, however, both indomethacin (10^{-6} M) and the more combined indomethacin (10^{-6} M) plus PGE_2 (10^{-7}-10^{-6} M) decreased apoptosis in MC of animals pretreated with indomethacin in vivo similar to that in PC. It is notable that PGE_2 effected apoptosis and changed cell viability in MC in a reverse concentration-dependent manner, and 10^{-8} M PGE_2 even increased apoptosis and decreased cell viability compared to indomethacin (10^{-6} M) alone in this time interval. These results suggest that during incubation, apparently in relation with changes in the other intracellular signal molecule concentrations, MC reduces its own sensitivity to PG apoptosis-modulating effects.

Several recent publications demonstrated that PGE_2 protects mammalian cells from apoptosis via the induction of cAMP synthesis [9, 25]. Our experiments demonstrated largely antiapoptotic effect of exogenous PGE_2 during incubation with MC, but not with PC, however, the similar effect was revealed in both fractions studied after 48 h. Such effect is likely due to the PGE_2-induced increase in cAMP levels similar to mammalian cells. However, effect of PGE_2 on cAMP synthesis may interfere with products following oxidant stress development during cell incubation [26], and this is a possible reason of change of proapoptotic effect of PG to antiapoptotic and inversely during incubation. In vertebrates, there was established a close relation between generation of reactive oxygen species (ROS) and the level of apoptosis in immune cells [4, 27]. In addition, involvement of reactive oxygen species in indomethacin-induced apoptosis was established for small intestinal epithelial cells [28].

To test the possibility of ROS influence on apoptosis-modulating effect of PGE_2, we used commercially available catalase in several experiments to diminish the ROS level. Interestingly, that addition of commercially available catalase together with PGE_2 (10^{-6} M) to MC pretreated with indomethacin in vivo did not influence decreasing effect of PGE_2 alone on apoptosis and cell viability in 24 h. In contrast, cell viability in PC administered with indomethacin in vivo was higher in experiments when PGE_2 (10^{-6} M) was used combined with catalase compared to addition of PGE_2 alone. It indicates that ROS indeed may interfere with PG effects on apoptosis in PC, but not in MC. Apparently, it can be explained by the higher proportion of phagocytes in PC compared to MC, and correspondingly higher oxidative activity of PC [18]. This agrees with the data on the relationship between functional activity and ROS generation in phagocytes of mammals [29]. Apparently, the lower sensitivity of holothurian MC to the proapoptotic effect of PGE_2 relative to phagocytes can be attributed to relatively low normal ROS generation in these cells.

As opposed to indomethacin, glucocorticoid hormones (GH) are physiological inhibitors of PG biosynthesis and are well known to be inductors of apoptosis in many cell types of vertebrates. In mature immune cells, one of the most often realized forms of apoptosis is that induced by GH [8]. Corticosteroid metabolites were found in sea stars [30], and hydrocortisone and corticosterone were found in several other seawater invertebrates [31]. Hydrocortisone was found to induce lysis in the center of the axial organ of sea stars *Asterias rubens* [32]. It was suggested that the cell death could occure via the apoptosis, as it is reminiscent of the effect of cortisone on the thymus of vertebrates.

In this chapter we have shown that synthetic GH dexamethasone also produced a concentration-dependent effect on the apoptosis level in normal coelomocyte fractions. In PC, dexamethasone (10^{-4}-10^{-7} M) induced apoptosis in the reverse concentration-dependent manner, and 10^{-4} M dexamethasone even inhibited apoptosis compared to the control in 24 h. However, dexamethasone (10^{-4} M) significantly increased apoptosis in P1 compared to the control at the same period of time. The different effects of dexamethasone in P1and PC indicate, apparently, that fractions of phagocytes P1 and P2 can cooperate during immune response, and resulting effect of dexamethasone may be mediated by humoral substances realizing by the cells. Data obtained previously on the increase in apoptosis level in P1 after addition the supernatants of P2 or MC pretreated with dexamethasone (10^{-4} M) compared to the antiapoptotic effects of supernatants of not-pretreated cells [33] susupport the idea.

In 48 h, dexamethasone (10^{-4} M) also inhibited apoptosis in MC compared to the control, and only the smallest concentration of dexamethasone of those studied (10^{-4}-10^{-7} M) decreased the cell viability.

Dexamethasone effects were also tested in fraction of phagocytes P1 after 48 h-incubation. As opposed to 24 h-time interval, there was shown reverse dexamethasone-concentration dependence of apoptosis level, and dexamethasone (10^{-4} M) inhibited apoptosis compared to the control. Thus, in 48 h, GH produced similar effects in P1 and MC.

Apparently, the inhibiting effect of 10^{-4} M dexamethasone on cell apoptosis is related to the feedback effect. This idea is supported by the fact that 10^{-4} M dexamethasone stimulated apoptosis in P1 in 24 h, but not in 48 h, and indicate that P1 is less sensitive to dexamethasone effect compared to PC.

Both indomethacin and dexamethasone (10^{-4} M) counteracted apoptosis-increasing effect of PG in normal MC, but failed to effect apoptosis reduction following PGE_2 (10^{-6} M) treatment in P2 in 48 h. Taken together, these data suggest that both of cyclooxygenase inhibitors studied may effectively inhibit PG biosynthesis in MC but not in phagocytes. In turn, in may be suggested that PG regulation of phagocyte activity has some special features compared to that in vertebrates. This idea is in line with the data reported by Pope et al. [34] on the lack of significant effect of indomethacin on the clearance of the bacteria from the coelomic cavity of starfish *Asterias rubens*. It is notable that coelomocytes of this starfish consist mostly of phagocyes [1].

Apparently, discrepancies between effects of PG regulation of immune response in vertebrates and echinoderms may be related to abnormally high oxidant activity of phagocytes of these marine invertebrates compared to that of vertebrates [18] . Expressed background apoptosis in the different fractions of freshly isolated coelomocyes of holothurian *E. fraudatrix* seems also to be the consequence of the high level of ROS generation in phagocytes.

The studies on catalase effect on apoptose-modulating effect of dexamethasone revealed that in normal P1, combined addition of dexamethasone (10^{-4} M) and catalase induced inhibition of apoptosis more pronounced compared to that induced by dexamethasone alone in 48 h. In normal PC, catalase together with dexamethasone (10^{-4} M) abolished apoptosis in 24 h. All these facts indicate that catalase benefited the decreasing effect of dexamethasone on apoptosis in fractions of phagocytes under study and increased cell viability in the cells where PGE_2 decreased it. However, no effects of catalase were shown on PGE_2 activity in MC in 24 h and on PGE_2- or dexamethasone (10^{-4} M)-modulating apoptosis in 48 h. These data correspond to those obtained for effects of catalase combined with indomethacin or indomethacin and PGE_2 in PC and MC discussed above and confirm the leading role of catalase in holothurian phagocyte protection from apoptosis.

On the whole, it is clear from our studies that PGE_2 may be involved in apoptosis modulation both in MC and phagocytes of holothurian *E. fraudatrix*. However, there are some differences in sensitivities of cell fractions studied to regulating effect of PG, and some peculiarities in effects of PGE_2 in phagocytes of holothurian *E. fraudatrix* and in vertebrates. The variations in the normal levels of ROS generation in different types of immune cells in accordance with the functional activities of cells and their environmental adaptation capacity, first of all to oxygen level, may at least partly contribute to these differences. Apparently, different sensitivities of MC and phagocyte fractions to apoptosis-modulating effect of PG may play an important role in cooperation of immunocytes during immune response similar to vertebrates. However, further studies are needed for clarification of the mechanisms of PGE_2 involvement in regulation of holothurian immunity.

REFERENCES

[1] Isaeva, VV; Korenbaum, ES. Protective functions of coelomocytes and immunity in Echinoderms. *Biol Morya.*, (Rus.). 1989, 6, 3-14.

[2] Chia, F; Xing, J. Echinoderm Coelomocytes. *Zool Stud.*, 1996, 35(4), 231-54.

[3] Eliseykina, MG; Magarlamov, Yu. Morphology of coelomocytes of holothurians *Apostichopus japonicus* (Aspidochirota: Stichopodidae) and *Cucumaria japonica* (Dendrochirota: Cucumariidae). *Biol Morya.*, (Rus.)., 2002, 28(3), 214-19.

[4] Zenkov, NK; Menshikova, EB; Volskii, NN; Kozlov, VA. Intracellular oxidative stress and apoposis. *Usp Sovrem Biol.*, (Rus.). 1999, 119(5), 440-50.

[5] Voronina, E; Wessel, G. Apoptosis in sea urchin oocytes, eggs and early embryos. *Mol Reproduct Develop.*, 2001, 60, 553-61.

[6] Rowley, AF; Vogan, CL; Taylor, GW; Clare, AS. Prostaglandins in non-insectan invertebrates: recent insights and unsolved problems. *J Exp Biol.*, 2005, 208, 3-14.

[7] Gromykhina, NY; Kozlov, V.A. Prostaglandin-dependent mechanisms of synthesis and action of immune control factors of macrophage and non-macrophage origin. *Immunologiya.*, 1996, 5, 29-34.

[8] McConkey, D; Orrenius, S; Jondal, M. Agents that elevate cAMP stimulate DNA fragmentation in thymocytes. *J Immunol.*, 1990, 145, 1227-30.

[9] Joseph, RR; Yazer, E; Hanakawa, Y; Stadnyk, AW. Prostaglandins and activation of AC/cAMP prevents anoikis in IEC-18. *Apoptosis.*, 2005, 10(6), 1221-33.

[10] Miller, JS; Nguyen, T; Stanley-Samuelson, DW. Eicosanoids mediate insect nodulation responses to bacterial infections. *Proc Natl Acad Sci* USA., 1994, 91, 12418-422.

[11] Stanley, D. Prostaglandins and other eicosanoids in insects: biological significance. *Annu Rev Entomol.*, 2006, 51, 25-44.

[12] Nomura, T; Ogata, H. Distribution of prostaglandins in the animal kingdom. *Biochem Biophys Acta.*, 1976, 431(1), 127-31.

[13] Mishchenko, TY; Korotchenko, OD; Isai, SV. Search for prostaglandin-like factors in marine organism. *Byull Eksp Biol Med.*, (Rus.). 1982, 94(11), 94-6.

[14] Levine, L; Kobayashi, T. Detection of compounds immunologically related to arachidonic acid transformation products in extracts of invertebrates. *Prostaglandins Leukot Med.*, 1983, 12(4), 357-69.

[15] Casterlin, ME; Reynolds, WW. Fever induced in marine arthropods by prostaglandin E1. *Life Sci.*, 1979, 25(18), 1601-03.

[16] Dolmatova, LS; Eliseykina, MG; Timchenko, NF; Kovaleva, AL; Shitkova, OA. Generation of reactive oxygen species in the different fractions of the coelomocytes of holothurian *Eupentacta fraudatrix* in response to the thermostable toxin of *Yersinia pseudotuberculosis* in vitro. *Chin J Limnol Oceanol.*, 2003, 21(4), 128-34.

[17] Bertheussen, K; Seljelid, R. Echinoid phagocytes in vitro. *Exp Cell Res.*, 1978, 111, 401- 12.

[18] Dolmatova, LS; Eliseikina, MG; Romashina, VV. Activity of antioxidant enzymes in

coelomocytes of *Eupentacta fraudatrix*. *Zh Evol Biokhim Fiziol*., (Rus.). 2004, 40(2), 104-11.

[19] Phillips, HJ. Dye exclusion tests for cell viability. In: Tissue culture methods and applications. New York: Academic *Press;* 1973, 406-408.

[20] Komatsu, N; Oda, T; Muramatsu, T. Involvement of both caspase-like proteases and serine proteases in apoptotic cell death induced by ricin, modeccin, diphtheria toxin, and *Pseudomonas* toxin. *J Biochem*., 1998, 124, 1038-44.

[21] Pramanick, D; Forstova, J; Pivec, L. 4 M guanidine hydrochloride applied to the isolation of DNA from different sources.*FEBS Lett*., 1976, 62(1), 81-4.

[22] Lee, RF; Steinert, S. Use of the single cell gel electrophoresis/comet assay for detecting DNA damage in aquatic (marine and freshwater) animals. *Mutat Res*., 2003, 544, 43-64.

[23] Romano, G; Russo, GL; Buttino, I; Ianora, A; Miralto, A. A marine diatom-derived aldehyde induces apoptosis in copepod and sea urchin embryos. *J Exp Biol*., 2003, 206, 3487-94.

[24] Knight, J; Taylor, GW; Wright, P; Clare, AS; Rowley, AF. Eicosanoid biosynthesis in an advanced deuterostomate invertebrate, the sea squirt (*Ciona intestinalis*). *Biochim Biophys Acta*., 1999, 1436, 467-78.

[25] Pino, MS; Nawrocki, ST; Cognetti, F; Abruzzese, JL; Xiong, HQ; McConkey, DJ. Prostaglandin E_2 drives cyclooxygenase-2 expression via cyclic AMP response element activation in human pancreatic cancer cells. *Cancer Biol Ther*., 2005, 4(11), 1263-69.

[26] Thelen, M; Dewald, B; Baggolini, M. Neutrophil sygnal transduction and activaton of the respiratory burst. *Physiol Rev*., 1993, 73 (4), 797-821.

[27] Davis, W; Ronae, Z; Tew, K. Cellular thiols and reactive oxygen species in drug-induced apoptosis. *J Pharmacol Exp Ther*., 2001, 296(1), 1-6.

[28] Omatsu, T; Naito, Y; Handa, O; Hayashi, N; Mizushima, K; Qin, Y; et al. Involvement of reactive oxygen species in indomethacin-induced apoptosis of small intestinal epithelial cells. *J Gastroenterol*., 2009, 44(Suppl 19), 30-34.

[29] Maianskiĭ, AN; Maianskiĭ, NA; Abadzhidi, MA; Zaslavskaia, MI. Apoptosis: the beginning of the future.*Zh Mikrobiol Epidemiol Immunobiol*., (Rus.). 1997, 2, 88-94.

[30] Hines, GA; Watts, SA; Sower, SA; Walker, CW. Sex steroid levels in the testes, ovaries and pyloric caeca during gametogenesis in the sea star *Asterias vulgaris*. *Gen Comp Endocrinol*., 1992, 87 (3), 451-60.

[31] Gurst, JE; Sheikh, YM; Djerassi, C. Synthesis of corticosteroids from marine sources. *J Am Chem Soc*., 1973, 95(2), 628-29.

[32] Leclerc, M; Bajelan, M. Effect of hydrocortisone on the axial organ of the sea star *Asterias rubens*. *Thymus*., 1991, 17(4), 253-56.

[33] Zaika, OA; Dolmatova, LS. Studies on possible mechanisms of cooperation of immunocytes of holothurian *Eupentacta fraudatrix* in immune response model. In: Proceedings of the 3d International Pacific congress on traditional medicine, 2006:Vladivostok, Seoul: *Vladivostok State Medical University*, 2006, 164-66.

[34] Pope, EC; Taylor, GW; Rowley, AF. Biosynthesis and functions of eicosanoids generated by the coelomocytes of the starfish, *Asterias rubens*. *Comp Biochem Physiol B Biochem Mol Biol*., 2007, 147(4), 657-66.

In: Prostaglandins: Biochemistry, Functions… ISBN: 978-1-61668-272-9
Editors: Gillian M. Goodwin, pp.65-80

Chapter 4

PHARMACOLOGICAL DIFFERENTIATION OF BIMATOPROST AND LATANOPROST INDUCED OCULAR HYPOTENSION BY A SECOND GENERATION PROSTAMIDE ANTAGONIST (AGN 211336)

F. David Woodward[1*], Yanbin Liang[1], W. Jenny Wang[1], Chen Li[1], M. Victor Guzman[1], B. Alexander Kharlamb[1], A. Larry Wheeler[1], E. Michael Garst[2], W. C. Robert Carling[3], L. Clive Cornell[3], G. Hans Fliri[3], L. Jose Martos[3], N. Simon Pettit[3], L. David Hopper[4], Vincenzo DiMarzo[5], and Giorgio Ortar[6]

[1]Department of Biological Sciences, Allergan Inc., Irvine, CA, USA
[2]Department of Chemical Sciences, Allergan, Inc., Irvine, CA, USA
[3]Selcia Ltd., Ongar, Essex, England
[4]Bioanalytical Systems Inc., Mount Vernon, IN, USA
[5]C.N.R., Pozzuoli, Naples, Italy
[6]Università di Roma 'La Sapienza', Rome, Italy

ABSTRACT

Previous studies using cells and tissue culture have demonstrated that prostamide and prostanoid FP receptor activities may be pharmacologically differentiated by selective prostamide receptor antagonists. The intent of the studies described herein was to reproduce this activity profile in the eyes of living animals. For this purpose, a second generation prostamide antagonist (AGN 211336) was used. AGN 211336 has a pA_2 of 7.6 at prostamide receptors but no meaningful activity (k_b>10,000 nM) at prostanoid FP and DP_1, DP_2, EP_{1-4}, and IP receptors. AGN 211336 also inhibited bimatoprost induced Cyr 61 mRNA upregulation in cells expressing the human recombinant prostamide

* Corresponding author: Department of Biological Science (RD3-2B), Allergan, Inc., 2525 Dupont Drive, Irvine, CA 92612, USA, 714-246-5490 (telephone), 714-246-5578 (fax), Woodward_david@allergan.com (email).

receptor but did not block the response to $PGF_{2\alpha}$. The effects of AGN 211336 on bimatoprost and latanoprost induced ocular hypotension and ocular surface hyperemia were evaluated in beagle dogs over a five day experimental period. Pretreatment with AGN 211336 (2x5% suspension daily) statistically significantly inhibited the decrease in intraocular pressure produced by bimatoprost (0.03% daily). In contrast, an identical dosing regimen for AGN 211336 did not alter the ocular hypotensive response to latanoprost (0.03% daily). Thus, the prostamide bimatoprost and the prostanoid FP receptor agonist prodrug latanoprost behaved as pharmacologically discrete entities in the eye, according to their differential susceptibility to AGN 211336. The ocular hypotensive effects of bimatoprost were not blocked by the FAAH inhibitor OL-92 or a close structural congener of AGN 211336 that lacked prostamide antagonist properties (AGN 220675). Interestingly, AGN 211336 did not block the ocular surface redness produced by either latanoprost or bimatoprost. This suggests that the ocular surface hyperemia produced by bimatoprost occurs as a result of non-prostamide, off-target activity.

Bimatoprost is a highly efficacious anti-glaucoma drug (Woodward et al., 2004). The underlying pharmacology of bimatoprost has been proposed to be that of a prostamide mimetic. This hypothesis was based on the differential pharmacology of bimatoprost and prostanoid FP receptor agonists in isolated tissues and cultured cells. Only certain isolated tissues were found responsive to prostamides such as bimatoprost: these tissues included the feline iris, lung parenchyma, and the rabbit uterus (Woodward et al., 2003; Matias et al., 2004; Chen et al., 2005a). On the other hand, numerous isolated preparations that constitutively expressed prostanoid FP receptors were essentially unresponsive to prostamides: these included the rat and gerbil colon, rabbit jugular vein, mouse ileum, and the human, rat, and mouse uterus (Woodward et al., 2003; Matias et al., 2004; Woodward et al., 2004; Chen et al., 2005a). Similarly, certain cells are also bimatoprost-sensitive (Liang et al., 2003; Spada et al., 2005), whereas cells natively expressing FP receptors and recombinant FP receptor transfectants are essentially unresponsive to bimatoprost (Woodward et al., 2003; Liang et al., 2004; Matias et al., 2004; Woodward et al., 2004; Spada et al., 2005). The prostamide receptor was thereby proposed as a distinct pharmacological entity based on this body of agonist data. The next step towards complete pharmacological characterization required the identification of an antagonist.

The first prostamide antagonist was described in 2007. This was AGN 204396. The isolated feline iridial preparation provided a useful system for discriminating prostamide and prostanoid FP receptor pharmacology since both activities co-exist in this tissue. AGN 204396 selectively antagonized the activity of the prostamides and bimatoprost in the feline iridial preparation but did not block the activity of prostaglandin (PG)$F_{2\alpha}$, its analogs 17-phenyl $PGF_{2\alpha}$, and latanoprost free acid (Woodward et al., 2007a). A similar activity profile was apparent for the antagonist congener AGN 204397 (Woodward et al., 2008a). Additionally, AGN 204396 and 204397 did not block native or recombinant FP receptors. These compounds enabled the prostamide receptor to be characterized as a singular entity at the pharmacological level. AGN 204396 and 204397 were, however, prototypes with pA_2 values in the 5.5 range and too weak for studies in living animals, where tissue levels and clearance cannot be controlled. More recently, prostamide antagonists 100 times more potent than AGN 204396 were discovered (Wan et al., 2007; Woodward et al., 2008b, Liang et al., 2008). One of these second generation and more potent prostamide antagonists (AGN

211334) was utilized to demonstrate that increases in conventional aqueous humor outflow produced by bimatoprost in the perfused human ocular anterior segment preparation were prostamide receptor mediated (Wan et al., 2007). A further compound in this series (AGN 211335) was used in structurally characterizing an FP/alt FP heterodimer as the human prostamide receptor (Liang et al., 2008). The purpose of these studies was to employ one of these second generation antagonists (AGN 211336) to reproduce, or otherwise, the differential pharmacology of a prostamide mimetic (bimatoprost) and a prostanoid FP receptor agonist latanoprost. The dog was selected for these ocular experiments since both bimatoprost and latanoprost profoundly lower intraocular pressure and cause a mild to moderate degree of conjunctival redness in this species.

METHODS

Isolated Feline Iris

Adult domestic cats were euthanized by i.v. overdose of sodium pentobarbital (Anthony, Arcadia, CA). The eyes were enucleated immediately thereafter and placed on ice. Two eyes provided a total of four iridial preparations. The iris sphincter was mounted vertically under 50 to 100 mg tension in a jacketed 10 ml organ bath. Smooth muscle tension of the isolated iris sphincter was measured isometrically with force displacement transducers (Grass FT-03) and recorded on a Grass polygraph (Model 7). The organ baths contained Krebs' solution maintained at 37°C by a heat exchanger and circulating pump. The Krebs' solution was gassed with 95% O_2, 5% CO_2 to give a pH of 7.4. and had the following composition: 118.0 mM NaCl, 4.7 mM KCl, 1.2 mM KH_2PO_4, 1.9 mM $CaCl_2$, 1.18 mM $MgSO_4$, 25.0 mM $NaHCO_3$, 11.7 mM glucose and 0.001 mM indomethacin. A 60 min stabilization period was provided before commencing each experiment. Activity was manifest as contractile responses and measured as such. The experiment was designed so that a direct, four-way comparison for antagonist vs. prostamide, vehicle vs. prostamide, antagonist vs. corresponding PG, and vehicle vs. corresponding PG was provided in tissue preparations obtained from a single animal. One cumulative dose-response curve to agonist was obtained in each tissue. Vehicle (ethanol) and antagonist (AGN 211336) were given 30 minutes before the agonist dose-response curves were constructed. The response to $PGF_{2\alpha}$ 10^{-7}M was determined at the beginning and end of each dose response curve, with appropriate wash-out, and responses were calculated as % of this reference contraction.

In order to calculate the pA_2 value for AGN 211336, the mean concentration response curve was plotted as log concentration response -1 (CR-1) vs log antagonist concentration, according to the method of Arunlakshana and Schild (1959) using GraphPad Prism 4 software. An adaptation of the prostamide $F_{2\alpha}$ concentration-response curve in the presence of vehicle was required for analysis. Thus, each point on the prostamide $F_{2\alpha}$ vs. vehicle concentration-response curve represents the mean of each of the six separate antagonist experiments. Since the slope of the CR-1 vs. log [Antagonist] plot did not significantly differ from unity, the slope was constrained to 1. The effects of antagonists were statistically analyzed by comparing the mid-points (E.C.$_{50}$s) of the agonist concentration-response curves in the presence or absence of antagonists using GraphPad Prism 4.

Ca^{2+} Signal Studies on Human Recombinant Prostanoid Receptors

The use of chimeric G protein cDNAs allowed responses to G_s and G_i coupled prostanoid receptors to be measured as a Ca^{2+} signal, as previously described (Woodward et al., 2003; Matias et al., 2004). Prostanoid DP, EP_2, and EP_4 receptor cDNAs were co-transfected with chimeric G_{qs} cDNA containing a haemagglutanin (HA) epitope. The prostanoid EP_3 receptor was co-transfected into HEK-293 EBNA cells, using $pCEP_4$ as a vector, with chimeric G_{qi}-HA. G_{qs} and G_{qi} chimeric cDNAs (Molecular Devices, Sunnyvale, CA) were cloned into a $pCEP_4$ vector and also selected by using a hygromycin B selection marker. Transfection into HEK-293 EBNA cells was achieved by the FuGENE 6 method. Because G_{qs} and G_{qi} contained an HA epitope, protein expression was detected by Western blotting analysis using anti-mouse HA monoclonal antibody and HRP conjugated secondary antibody. For human recombinant EP_1, FP, IP, and TP receptors, stable transfectants were obtained as previously described (Woodward et al., 2003; Matias et al., 2004). Briefly, $pCEP_4$ was used as the expression vector and transfection into HEK-293-EBNA cells was performed with FuGENE 6. Stable transfectants were again selected according to hygromycin resistance.

Ca^{2+} signaling studies were performed using a FLIPR (fluorometric imaging plate reader) instrument. Cells were seeded at a density of 5 x 10^4 cells/well in Biocoat poly-D-lysine coated, black wall, clear bottom 96 well plates (BD Biosciences, Franklin Lakes, NJ) and allowed to attach overnight in an incubator at 37°C. The cells were then washed twice with HBSS-HEPES buffer (Hanks' balanced salt solution without bicarbonate and phenol red, 20 mM HEPES, pH 7.4) using a Denley Cellwash plate washer (Labsystems, Franklin, MA). After 45-60 min of dye-loading in the dark using the Ca^{2+}-sensitive dye Fluo-4AM, at a final concentration of 2 x 10^{-6}M, the plates were washed 4 times with HBSS-HEPES buffer to remove excess dye and leaving 100μl of buffer in each well. The plates were then placed in the FLIPR instrument and allowed to equilibrate at 37°C. Compound solutions were added in a 50μl volume to each well to give the desired final concentration. Cells were excited with an argon laser at 488 nm and emission was measured through a 510-570 nm band width emission filter (FLIPR, Molecular Devices, Sunnyvale, CA). The peak increase in fluorescence intensity was recorded for each well. On each plate, four wells each served as negative (HBSS-HEPES buffer) and positive controls (standard agonist: DP = BW 245C, EP_1-EP_4 = PGE_2, FP = $PGF_{2\alpha}$, IP = carbaprostacyclin TP=U-46619). The peak fluorescence change in each well containing drug was expressed relative to the controls. To obtain concentration-response curves, compounds were tested in duplicate in each plate over the desired concentration range. Each compound was tested on at least 3 separate plates using cells from different passages to give n=3.

Human Recombinant Prostamide Receptor Studies

The prostamide receptor was recently described as an FP/alt FP heterodimer (Liang et al., 2008) and cell lines co-expressing FP and alt FP were prepared as follows. FP and alt FP cDNAs were subcloned into a retrovirus vector (Invitrogen). The plasmids were then transfected into GP2-293 packaging cells. The supernatants containing virus were harvested after 72hr and cell debris was removed by centrifugation (800g x 15min). HEK 293 / EBNA cells were co-infected with alt FP and FP wild type in equal amounts of virus.

Ocular Studies

Male and female dogs of at least seven months of age were used for the ocular studies. Intraocular pressure was measured in conscious dogs by pneumatonometry (Model 30 classic pneumatomometer, Medtronics, MA). Ocular surface hyperemia was visually assessed and scored according to a clinical grading scale (0.5 = trace, 1.0 = mild, 2.0 = moderate, 3.0 = severe), with the aid of photographs for reference.

The experimental design involved groups of dogs comprised of four female and four male dogs. The drugs were coded as drug A (AGN 211336 or vehicle), drug B (AGN 211336 or vehicle), or drug C (bimatoprost 0.03% in one study, latanoprost 0.03% in the other study). Experiments on OL-92 and AGN 220675 were designed in an identical manner. At time 0, the intraocular pressure and ocular surface hyperemia were measured and assessed, respectively. Drug A was administered to one eye and drug B administered to the contralateral eye immediately afterwards. Two hours later (time 2hr) intraocular pressure and ocular surface hyperemia were again measured / assessed and drugs A and B were administered for a second time. Drug C (bimatoprost or latanoprost) was then topically applied to both eyes. Intraocular pressure was measured and ocular surface hyperemia assessed and scored two hours (time 4 hr) and four hours (time 6 hr) post-drug C application. This daily regimen was repeated over the next four consecutive days to result in a 5 day study. The dog studies were performed in a masked fashion such that the experimenters were unaware of the identities of drugs coded A, B, and C. The dog studies were performed at Bioanalytical Sciences Inc. (Evansville, IN, USA).

The data were analyzed as the decrease after the second AGN 211336 (OL-92 or AGN 220675) dose on day 1. The data were statistically analyzed by Students' paired t test. Dogs that exhibited an intraocular pressure of less than 13 mmHg in one or both eyes or hyperemia with a score of one difference between the two eyes at the 2hr time point on day 1 were removed from the statistical analyses.

AGN 211336, OL-92, and AGN 220675 were suspended in peanut oil at 5%. Bimatoprost 0.03% and latanoprost 0.03% were prepared in 1% polysorbate/10mM TRIS-HCl. Nepafenac (Alcon Laboratories, Fort Worth, TX) was purchased as the commercially available formulation, 0.1% suspension. AGN 211336 was designed and synthesized at Selcia Ltd., Ongar, Essex, England.. Bimatoprost was supplied by the Allergan Pharmaceuticals Science Department. Prostaglandins D_2, E_2, and $F_{2\alpha}$, U-46619, BW245C, AL-8810 and latanoprost were purchased from Cayman Chemical (Ann Arbor, MI). Prostamide $F_{2\alpha}$ was synthesized at Allergan, Inc. or purchased from Cayman Chemical, Kalamazoo, MI, USA. OL-92 was synthesized by the laboratory of Prof. G. Ortar (Università di Roma, Italy). AGN 220675 was synthesized by Target Molecules. (Southampton, England).

RESULTS

AGN 211336 was characterized as a selective prostamide antagonist: these results are summarized in Table 1. AGN 211336 blocked the effect of bimatoprost in the isolated feline iris sphincter preparation with a pA_2 of 7.6 but was inactive vs. $PGE_{2\alpha}$. In HEK-293 stably expressing each of the human recombinant prostanoid receptors, AGN 211336 was inactive in preventing the Ca^{2+} signal response at DP_1, EP_1, EP_2, EP_3, EP_4, FP and IP receptors; it was, however, a TP antagonist (K_i = 111nM). In HEK-293 cells stably expressing the prostamide receptor by co-transfecting FP and alt. FP4, AGN 211336 dose-dependability inhibited bimatoprost induced Cyr 61 mRNA upregulation but did not affect the response to $PGF_{2\alpha}$ (Figure 1). The pharmacology of AGN 220675 is summarized in Table 2: no meaningful activity at prostanoid receptors or the prostamide receptor was apparent.

Table 1. Antagonist Activity of AGN 211336 at Prostamide Receptors in the Feline Iris and at Human Recombinant Prostanoid Receptors

Prostanoid Receptor	Antagonist K_b (nM) or pA_2 for AGN 211336
Prostamide (feline iris)	7.6 (pA_2)
DP_1	40861 (nM)
EP_1	Inactive
EP_2	Inactive
EP_3	Inactive
EP_4	Inactive
FP	Inactive
IP	Inactive
TP	111 (nM)

Table 2. Antagonist Activity of AGN 220675 at Prostamide Receptors in the Feline Iris and at Human Recombinant Prostanoid Receptors.

Prostanoid Receptor	Antagonist Kb (nM) or pA2 for AGN 220675
Prostamide (feline iris)	Inactive
DP1	Inactive
EP1	Inactive
EP2	Inactive
EP3	Inactive
EP4	3957
FP	5932
IP	Inactive
TP	68 (nM)

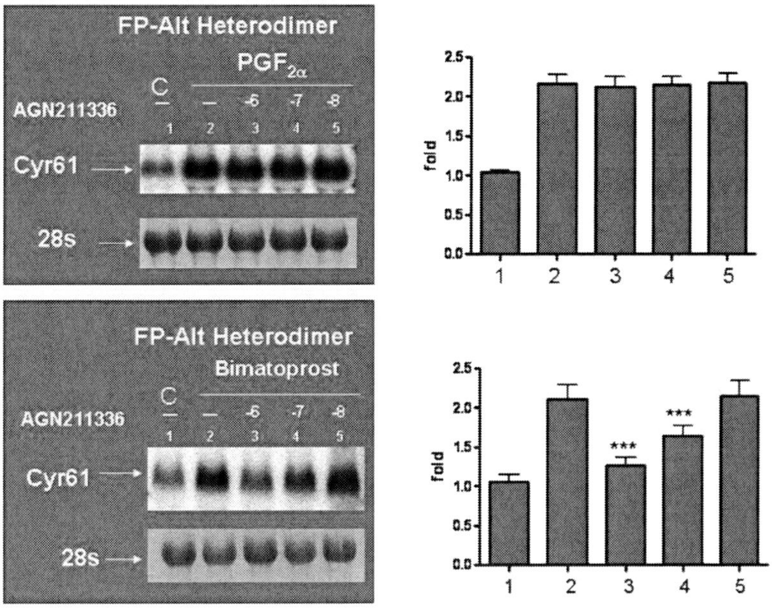

Figure 1. Northern Blot Analysis of Cry61 Gene Expression Following $PGF_{2\alpha}$ and Bimatoprost Treatment. HEK293/EBNA cells co-expressing FP and altFP4 receptors were pretreated with 10^{-6} to 10^{-8}M AGN 211336 for 15 min (*top and bottom panels, lane 3 to 5*) and treated with 10^{-7}M $PGF_{2\alpha}$ for 1 hr (*top panel, lane 2 to 5*) or treated with 10^{-7}M bimatoprost for 1hr (*bottom panel, lane 2 to 5*). *Arrows* indicate Cyr61 mRNA levels. The data represent mean ± S.D. of three independent densitometric experiments, ***$p<0.01$ *versus* bimatoprost alone (*lane 2*). The intensities of the 28S rRNA bands were used to normalize the RNA loading differences.

The effect of the prostamide antagonist AGN 211336 on Bimatoprost induced ocular hypotension in dogs is graphically depicted in Figure 2. Daily treatment with AGN 211336 significantly reduced the ocular hypotensive response to bimatoprost at most time points. In the presence of AGN 211336 (5%) the decrease in intraocular pressure was maintained at about 2-3 mmHg whereas in vehicle treated eyes intraocular pressure steadily decreased to approach 7 mmHg over the five day experimental period (Figure 2).

The effect of the prostamide antagonist AGN 211336 on latanoprost induced ocular hypotension is shown in Figure 3. Once daily dosing with 0.03% latanoprost caused decreases in intraocular pressure that peaked at about 6-7 mmHg. The effect of latanoprost on intraocular pressure was not altered by daily AGN 211336 (5%) treatment, and no significant differences occurred. AGN 220675, a close structural congener of AGN 211336 where the isopropylamide group is replaced by serinolamide to render the molecule inactive as a prostamide antagonist, did not block the effect of bimatoprost on dog intraocular pressure (Figure 4). A potent FAAH inhibitor, OL-92, (Lichtman et al., 2004; Boger et al., 2005) did not affect the ocular hypotensive activity of bimatoprost (Figure 5). AGN 220675 and OL-92 did not affect the ocular surface hyperemic effect of bimatoprost (data not shown). Nepafenac, an amido COX inhibitor claimed to be converted to free acid derivative armfenac (Maxwell et al., 2008) did not reverse the ocular hypotensive effect of bimatoprost (Figure 6). There were several instances where nepafenac actually significantly increased the activity of bimatoprost; significant inhibitor occurred at one time point. There is no obvious explanation

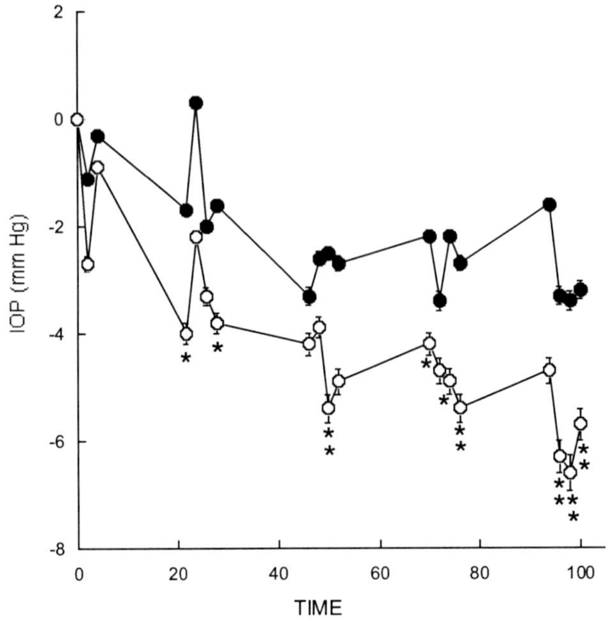

Figure 2. Effect of a prostamide antagonist AGN 211336 on bimatoprost induced ocular hypotension in beagle dogs. Eyes that received AGN 211336 (5%) pretreatment before bimatoprost (0.03%) are represented by the filled circles; eyes that received vehicle pretreatment before bimatoprost (0.03%) are represented by open circles. Values are mean ± S.E.M.; n = 6; *P<0.05; **P<0.01.

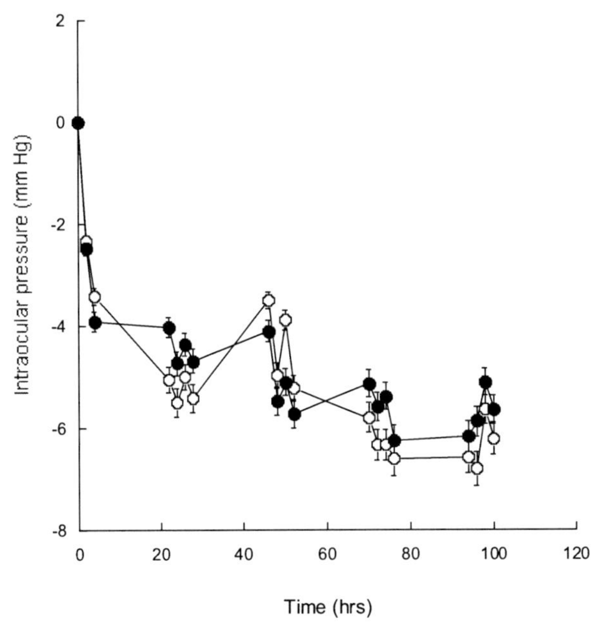

Figure 3. Effect of a prostamide antagonist AGN 211336 on latanoprost induced ocular hypotension in beagle dogs. Eyes that received AGN 211336 (5%) pretreatment before latanoprost (0.03%) are represented by filled circles, eyes that received vehicle pretreatment before latanoprost (0.03%) are represented by open circles. Values are mean ± S.E.M.; n = 6.

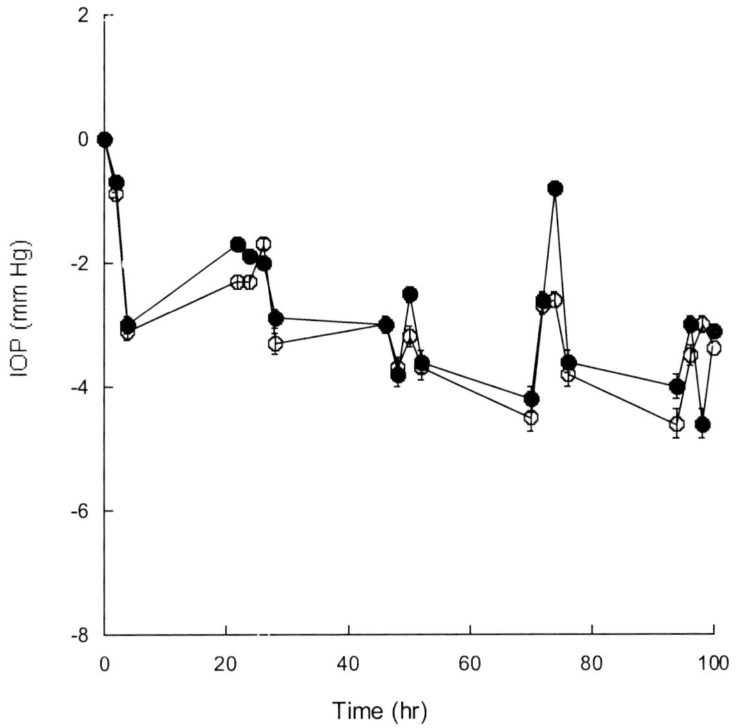

Figure 4. Effect of AGN 220675 (5%), a structural congener of AGN 211336 but lacking prostamide antagonist activity, on bimatoprost induced ocular hypotension in beagle dogs. Eyes that received AGN 220675 (5%) pretreatment before bimatoprost (0.03%) are represented by the filled circles, eyes that received vehicle pretreatment before bimatoprost (0.03%) are represented by open circles. Values are mean ± S.E.M.; n=5, *P<0.05.

for this result but it does indicate that (a) bimatoprost effects on intraocular pressure are not mediated by secondary PG release (b) bimatoprost does not compete with nefemac for ocular hydrolases. In marked contrast to intraocular pressure, AGN 211336 did not meaningfully alter the ocular surface hyperemia produced by bimatoprost (Figure 8). Similarly AGN 211336 did not alter the ocular surface redness caused by latanoprost (Figure 9.). No significant differences occurred at any time point.

Finally, the structures of AGN 211336, AGN 220675 and OL-92 are depicted in Figure 8. Note the structural similarity of AGN 211336 and AGN 220675, substitution of isopropylamide by serinolamide results in loss of activity at the prostamide receptor (see tables 1 and 2).

Discussion

AGN 211336 is a second generation prostamide antagonist. It has a pA_2 value of 7.6 at the feline prostamide receptor and no activity at prostanoid DP, $EP_{1\text{-}4}$, FP or IP receptors at a 30,000nM concentration. It had a k_d value of 111nM at the TP receptor but this does not interfere with the *in vivo* prostamide and prostanoid FP receptor studies described herein. The prostamide receptor has recently been defined as a heterodimeric combination of FP and an alternative mRNA splicing variant alt FP4 (Liang et al., 2008) and AGN 211336 dose-dependently antagonized bimatoprost induced Cyr 61 mRNA upregulation in the human recombinant prostamide preparation. AGN 211336 was, therefore, considered sufficiently potent and efficacious to be a useful compound for in vivo studies. AGN 211336 was used to determine whether prostamide pharmacology determined in vitro would transition into living animal studies, specifically ophthalmology studies.

AGN 211336 was indeed sufficiently potent to block the effect of bimatoprost on intraocular pressure. The effects of AGN 211336 were long-acting, with the ocular hypotensive response to bimatoprost held to no more than 3mmHg over days 2-5. In control eyes, bimatoprost was increasingly effective and intraocular pressure gradually decreased to about 7 mmHg. AGN 211336 is almost certainly resident in ocular tissues throughout the experimental time course for the following reasons. Firstly, it is very lipoplilic and would likely have a long residence time. Secondly, the semi-permanent changes in the ciliary muscle produced by bimatoprost (Richter et al., 2003) would likely remain largely unaffected by an antagonist with a short residence time by virtue of breakthrough phenonmenon.

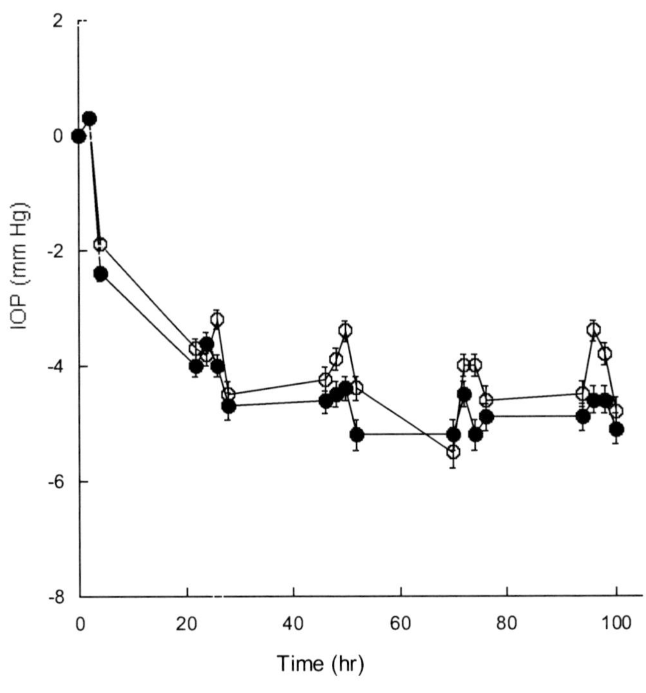

Figure 5. Effect of OL-92 (5%), an FAAH inhibitor, on bimatoprost induced ocular hypotension in beagle dogs. Eyes that received OL-92 (5%) pretreatment before bimatoprost (0.03%) are represented by the filled circles, eyes that received vehicle pretreatment before bimatoprost (0.03%) are represented by open circles. Values are mean ± S.E.M; n = 6; **P<0.01.

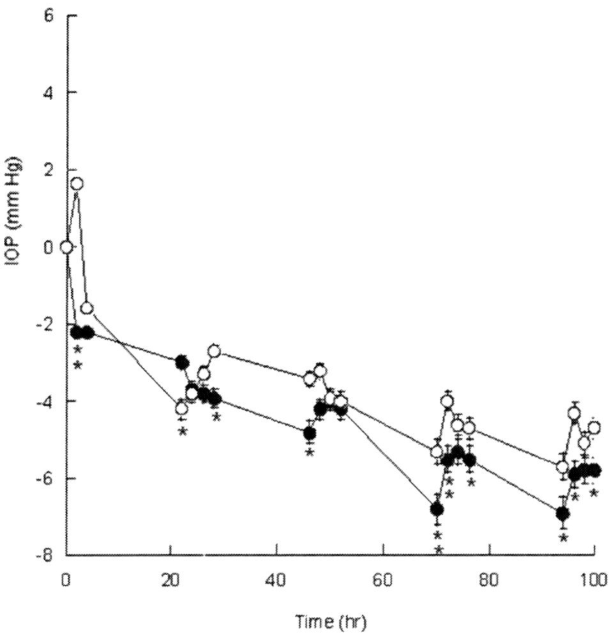

Figure 6. Effect of the COX inhibitor nepafenac (0.1%) on bimatoprost induced ocular hypotension in beagle dogs. Eyes that received nepafenac (0.1%) pretreatment before bimatoprost (0.03%) are represented by the filled circles, eyes that received vehicle pretreatment before bimatoprost (0.03%) are represented by open circles. Values are mean ± S.E.M.; n=5.

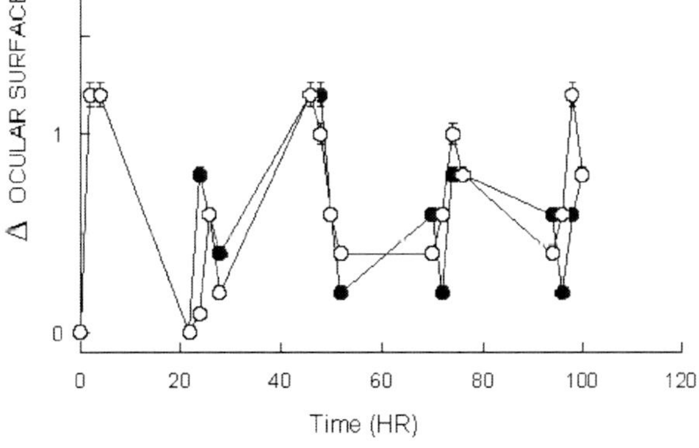

Figure 7. Effect of a prostamide antagonist AGN 211336 on bimatoprost induced ocular surface hyperemia in beagle dogs. Eyes that received AGN 211336 (5%) pretreatment before bimatoprost (0.03%) are represented by filled circles; eyes that received vehicle pretreatment before bimatoprost (0.03%) are represented by open circles. Values are mean ± S.E.M.; n = 5; *P<0.05.

AGN 211336 did not block the ocular hypotensive effect of the FP receptor agonist prodrug latanoprost. Both bimatoprost and latanoprost were given at an identical dose (0.03%). They also produced a similar degree of ocular hypotensive activity. These results suggest that the ocular hypotensive effects of latanoprost are not prostamide receptor mediated. Attempts to provide a positive control using the purported prostanoid FP receptor antagonist AL-8810 (Griffin et al., 1999) were compromised by its pronounced ocular hypotensive activity and extreme irritant effects in the dog eye (data not shown, experiment was terminated early and the eyes washed out). The activity of AGN 211336 was confirmed indirectly as prostamide-specific by the lack of effect of a close structural congener, which lacked activity at prostamide receptors (AGN 200675). The role of FAAH activity was also investigated. The highly potent FAAH inhibitor

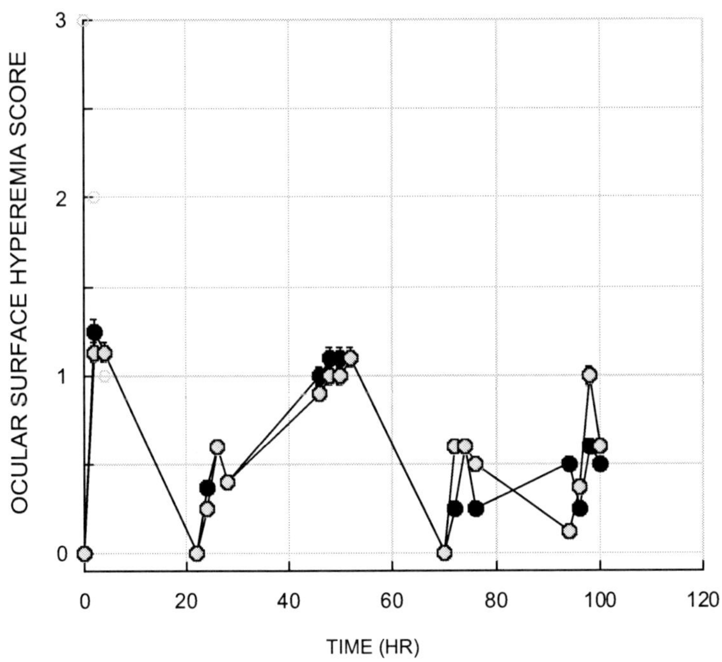

Figure 8. Effect of a prostamide antagonist AGN 211336 on latanoprost induced ocular surface hyperemia in beagle dogs. Eyes that received AGN 211336 (5%) pretreatment before latanoprost (0.03%) are represented by filled circles; eyes that received vehicle pretreatment before latanoprost (0.03%) are represented by open circles. Values are mean ± S.E.M.; n = 6.

OL-92 (Lichtman et al.; Boger et al., 2005), did not alter the effects of bimatoprost in the dog eye. Since bimatoprost has also been demonstrated inactive as an FAAH substrate (Matias et al., 2004), these studies demonstrate a lack of FAAH involvement in ocular prostamide pharmacology.

Clinical studies in patients refractory to the ocular hypotensive effects of the FP receptor agonist prodrug latanoprost have indicated that bimatoprost has a distinct pharmacology (Gandolfi and Cimino, 2003). This present ophthalmological study extends and confirms the earlier findings. The prostamide antagonist AGN 211336 blocks the ocular hypotensive effects of bimatoprost but not those of latanoprost. This provides further evidence that FP receptors and prostamide-sensitive receptors exist as discrete entities in the eye.

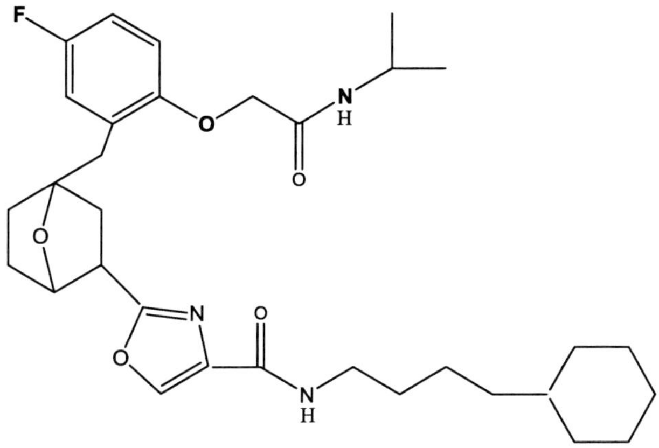

AGN 211336

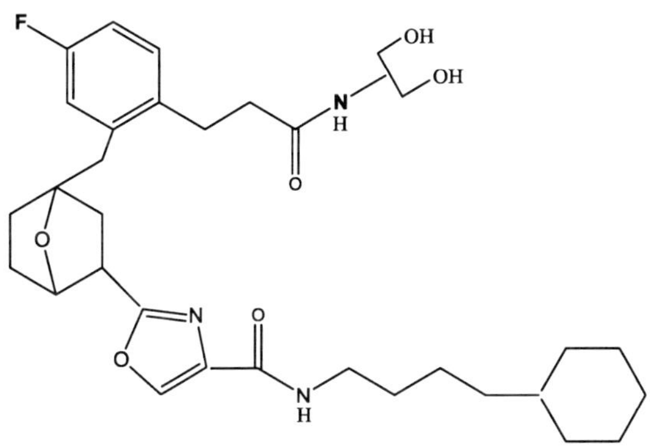

AGN 220675

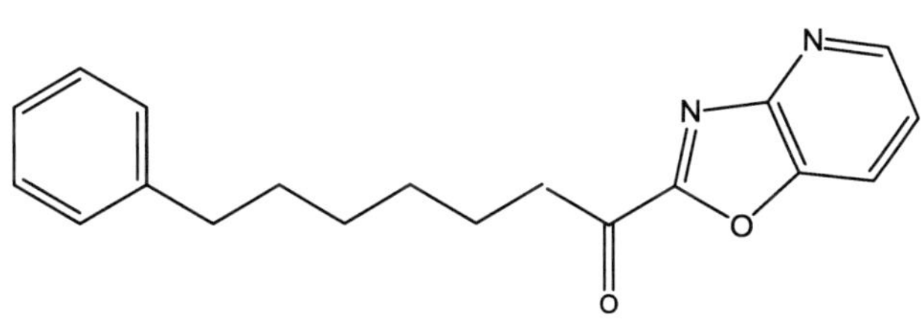

OL - 92

Figure 9.

AGN 211336 did not alter the ocular surface hyperemia produced by either latanoprost or bimatoprost. It is hardly surprising that AGN 211336 did not alter latanoprost induced ocular surface redness. An FP receptor associated with the vascular endothelium has been reported (Chen et al., 1995) and the resultant vasodilatation does not occur in response to prostamides (Chen et al., 2005b). AGN 211336 hardly affected bimatoprost induced ocular surface hyperemia, despite clearly and substantially blocking the ocular hypotensive response. These results suggest that prostamide receptor activation produces little ocular surface vasodilatation. This leaves the question, how does bimatoprost produce ocular surface hyperemia at the pharmacological level. Studies in monkeys have shown that bimatoprost is present in ocular surface tissues at high concentrations, 1-3 x 10^{-5}M (Woodward et al., 2003). Such a concentration may result in off-target effects on ocular surface blood vessels. Previous studies have clearly demonstrated that bimatoprost induced ocular surface hyperemia is physiological in that it is blocked by an NOS inhibitor, similar to latanoprost (Chen et al., 2005). This raises the possibility that bimatoprost may be at sufficiently high levels to weakly activate anandamide-sensitive and certain prostanoid receptors (Woodward et al., 2003; Matias et al., 2004), which results in observable conjunctival redness. In monkey eyelids and conjunctival tissue, conversion of bimatoprost to the FP receptor agonist 17-phenyl $PGF_{2\alpha}$ is minimal to absent (Woodward et al., 2003). It is unlikely therefore, that the ocular surface hyperemic effects are the result of enzymatic hydrolysis of bimatoprost.

In summary, the prostamide antagonist AGN 211336 provided a "tool" drug that allowed prostamide and prostanoid FP receptor agonist effects on ocular hypotension to be pharmacologically discriminated. An absence of FAAH activity with respect to both prostamide agonist and antagonist effects in the eye was further established. The reported loss of bimatoprost ocular hypotensive activity in FP receptor - / - mice (Crowston et al., 2005; Ota et al., 2005) is consistent with the prostamide receptors as an FP/alt FP4 heterodimeric complex (Liang et al., 2008), since both would be encoded by the FP receptor gene. This study provides further evidence for the prostamide receptor as a discrete entity in the eye.

Abbreviations

PG	(prostaglandin)
FP	(prostanoid FP receptor subtype)
alt FP4	(prostanoid FP receptor alterative mRNA splicing variant 4)
FAAH	(fatty acid amide hydrolase)
HEK	(Human embryonic kidney cells)
IOP	(intraocular pressure)

References

Arunlakshana, O. & Schild, H. O. (1959). Some quantitative use of drug antagonists. *Br. J. Pharmacol. Chemother.*, *257*, 48-58.

Boger, D. L., Miyauchi, H., Du, W., Hardovin, C., Fecik, R. A., Cheng, H., Hwang, I., Hendrick, M. P., Leung, D., Acevedo, O., Guimaraes, C. R. W., Jargensen, W. L. & Cravatt, B. F. (2005). Discovery of a potent, selective, and efficacious class of reversible α-ketoheterocycle inhibitors of fatty acid amide hydrolase effective as analgesis. *J. Med. Chem.*, *48*, 1849-1856.

Chen, J., Champa-Rodriguez M. L. & Woodward, D. F. (1995). Identification of a Prostanoid FP receptor population producing endothelium-dependent vasorelaxation in the rabbit jugular vein. *Br. J. Pharmacol.*, *116*, 3035-3041.

Chen, J., Dinh, T., Woodward, D. F., Holland, J. M., Yuan, Y. D., Lin T. H. & Wheeler, L. A. (2005). Bimatoprost: mechanism of ocular surface hyperemia associated with topical therapy. *Cardovasc. Drug Rev.*, *23*, 231-246.

Chen, J., Senior, J., Marshall, K., Abbass, F., Dinh, H., Dinh, T., Wheeler, L. A. & Woodward, D. F. (2005a). Studies using isolated uterine and other preparations show bimatoprost and prostanoid FP agonists have different activity profiles. Br. *J. Pharmcol.*, *144*, 493-501.

Chen, J., Dinh, T., Woodward, D. F., Holland, J. M., Yuan, Y. D., Lin, T. H. & Wheeler, L. A. (2005b). Bimatprost: mechanism of ocular surface hyperemia associated with topical therapy. Cardiovasc. *Drug Rev.*, *23*, 231-246.

Crouston, J. G., Lindsey, J. D., Morris, C. A., Wheeler, L., Medeinos, F. A. & Weinreb, R. N. (2005). Effect of bimatoprost on intraocular pressure in prostaglandin FP receptor knockout mice. *Invest. Ophthalmol. Vis. Sci.*, *46*, 4571-4577.

Gandolfi , S. A. & Cimino, L. (2003). Effect of bimatoprost on patients with primary open-angle glaucoma on ocular hypertension who are nonresponders to latanoprost. *Ophthalmol.*, *110*, 609-614.

Griffin, B. W., Klimko, P., Crider, J. Y. & Sharif, N. A. (1999). AL-8810: a novel prostaglandin $F_{2\alpha}$ analog with selective antagonist effects at the prostaglandin $F_{2\alpha}$ (FP) receptor. *J. Pharmacol. Exp. Ther.*, *290*, 1278-1284.

Liang, Y., Li, C., Guzman, V. M., Evinger, A. J., Protzman, C. E., Krauss, A. H. P. & Woodward, D. F. (2003). Comparison of $PGF_{2\alpha}$, bimatoprost (prostamide) and butaprost (EP_2 agonist) on Cyr 61 and CTGF gene expression. *J. Biol. Chem.*, *278*, 27267-27277.

Liang, Y., Li., C., Guzman, V. M., Chang, W. W., Evinger, A. & Pablo, J. V. Woodward, D. F. (2004). Upregulation of orphan nuclear receptor Nur 77 following $PGF_{2\alpha}$, bimatoprost and butaprost treatments. Essential role of a protein kinase C pathway involved in the EP_2 receptor activated Nur 77 gene transcription. Br. *J. Pharmacol.*, *142*, 737-748.

Liang, Y., Woodward, D. F., Guzman, V. M., Li, C., Scott, D. F., Wang, J. W., Wheeler, L. A., Garst, M. E., Landsverk, K., Sachs, G., Krauss, A. H. P., Cornell, C., Martos, J., Pettit, S. & Fliri, H. (2008). Isolation and pharmacological characterization of the prostaglandin FP receptor and FP receptor variant complexes. Br. *J. Pharmacol.*, *154*, 1079-1093.

Lichtman, A. H., Leung, D., Shelton, C., Saghatelian, A., Hardovin, C., Boger, D. L. & Cravatt, B. F. (2004). Reversible inhibitors of fatty acid amide hydrolase that promote analgesia: evidence for an unprecedented combination of potency and selectivity. *J. Pharmacol. Exp. Ther.*, *311*, 441-448.

Matias, I., Chen, J., De Petrocellis, L., Bisogno, T., Ligresti, A.,Fezza, F., Krauss, A. H. P. Shi, L., Protzman, C. E., Li, C., Liang, Y., Nieves, A. L., Kedzie, K. M., Burk, R. M., Di Marzo, V. & Woodward, D. F. (2004). Prostaglandin ethanolamides (prostamides) = *in vitro* pharmacology and metabolism. *J. Pharmacol. Exp. Ther.*, *309*, 745-757.

Maxwell, W. A., Reiser, H. J., Stewart, R. H., Cavanagh, H. D., Walters, T. R., Sager, D. P. & Meuse, P. A. (2008). the Nepafenac Study Group Nepafenac dosing frequency for ocular pain and inflammation associated with cataract surgery. *J. Ocular Pharmacol., Ther.*, *24*, 593-599.

Ota, T., Aihara, M., Narumiya, S. & Araie, M. (2005). The effects of prostaglandin analogues on IOP in prostanoid FP receptor-deficient mice. *Invest. Ophthalmol., Vis. Sci., 46*, 4159-4163.

Richter, M., Krauss, A. H., Woodward, D. F. & Lutjen-Drecoll, E. (2003). Morphological changes in the anterior eye segment after long-term treatment with different receptor selective prostaglandin agonists and a prostamide. Invest. Ophthalmol. *Vis. Sci., 44*, 4419-4426.

Spada, C. S., Krauss, A. H. P., Woodward, D. F., Chen, J., Protzman, C. E., Nieves, A. L., Wheeler, L. A., Scott, D. F. & Sachs, G. (2005). Bimatoprost and prostaglandin $F_{2\alpha}$ selectively stimulate intracellular calcium signaling in different cat iris sphincter cells. *Exp. Eye Res., 80*, 135-145.

Wan, Z., Woodward, D. F., Cornell, C. L., Fliri, H. G., Martos, J. L., Pettit, S. N., Wang, J. W., Kharlamb, A. B., Wheeler, L. A., Garst, M. E., Landsverk, K. J. Struble, C. S. & Stamer, W. D. (2007). Bimatoprost , prostamide activity , and conventional drainage. *Invest. Ophthalmol. Vis. Sci., 48*, 4107-4115.

Woodward, D. F., Krauss, A. H. P., Chen, J., Liang, Y., Li, C., Protzman, C.E., Bogardus, A., Chen, R., Kedzie, K. M., Krauss, H. A., Gil, D. W., Kharlamb, A., Wheeler, L. A., Babusis, D., Welty, D., Tang-Liu, D. D. S., Cherkury, M., Andrews, S. W., Burk, R. M. & Garst, M. E. (2003). Pharmacological characterizations of a novel antiglaucoma agent, bimatoprost (AGN 192024). *J. Pharmacol. Exp. Ther., 305*, 772-785.

Woodward, D. F., Phelps, R. L., Krauss, A. H. P., Weber, A., Short, B., Chen, J., Liang, Y., Wheeler, L. A. (2004). Bimatoprost: a novel anti-glaucoma agent. Cardiovasc. *Drug Rev., 22*, 103-120.

Woodward, D. F., Krauss, A. H. P., Wang, J. W., Protzman, C. E., Nieves, A. L., Liang, Y., Donde, Y., Burk, R. M., Landsverk, K. & Struble, C. (2007a). Br. *J. Pharmacol., 150*, 342-352.

Woodward, D. F., Liang, Y. & Krauss, A. H. P. (2008). Prostamides (prostaglandin-ethanolamides) and their pharmacology. Br. *J. Pharmacol., 153*, 410-419.

Woodward, D. F., Carling, R. W. C., Cornell, C. L. Fliri H. G. , Martos, J. L., Liang, Y. & Wang J. W. (2008b). The Pharmacology and Therapeutic relecvance of endocannabinoid derived cyclo-oxygenase (COX) -2 products. *Pharmacol.*, Ther , in press.

In: Prostaglandins: Biochemistry, Functions…
Editors: Gillian M. Goodwin, pp.81-92 ISBN: 978-1-61668-272-9

Chapter 5

Regulation of Myeloid Derived Phagocytes by Phosphatidylserine-Containing Liposomes: Possible Involvement of Prostaglandin E_2 and the Potential Therapeutic Application

Zhou Wu* and Hiroshi Nakanishi*
Department of Aging Science and Pharmacology, Faculty of Dental Science,
Kyushu University, Fukuoka 812-8582, Japan

Abstract

Phosphatidylserine (PS), which is normally located on the inner leaflet of the plasma membrane, translocates to the outer leaflet during early stages of apoptosis,. The externalization of PS provides a signal for phagocytes to initiate uptake of apoptotic cells. After the phagocytosis of apoptotic cells, phagocytes start to secrete anti-inflammatory mediators including prostaglandin E_2 (PGE_2), which act an autocrine manner to control their functions. PS-containing liposomes (PS liposomes) can mimic the effects of apoptotic cells on phagocytes to induce the secretion of PGE_2. PS liposomes induced PGE_2 secretion from microglia without induction of either cyclooxygenase (COX)-2 or microsomal prostaglandin E synthase (mPGES)-1. PS liposomes were found to utilize the COX-1/mPGES-2 system to produce PGE_2 secretion and then shift microglia and macrophages from a pro- to anti-inflammatory phenotype by the autocrine action of PGE_2. Moreover, PS liposomes inhibit the maturation of dendritic cells and osteoclast precursors. Therefore, PS liposomes may be potentially useful as one type of pharmacological intervention for inflammatory and immune diseases through a feedback mechanism involving PGE_2.

* Corresponding author: Department of Aging Science and Pharmacology, Faculty of Dental Science, Kyushu University, Fukuoka 812-8582, Japan. Fax: 81-91-642-6415, E-mail: zhouw@dent.kyushu-u.ac.jp (ZW) and nakan@dent.kyushu-u.ac.jp (HN)

INTRODUCTION

Mammalian myeloid derived mononuclear cells including monocytes in the blood, macrophages in the spleen and microglia in the brain, can phagocytoze apoptotic cells and necrotic cell debris in the tissues. These cells are known as phagocytes [Athanasou et al., 1990]. In addition to these professional phagocytes, recent reports have also shown that non-professional phagocytes, such as dendritic cells, the myeloid-derived professional antigen presenting cells (APC), can phagocytoze apoptotic cells [Steinman et al., 2000; Iyoda et al., 2002]. Moreover, the myeloid derived multinucleated cells, osteoclasts, which generally remove mineralized matrix resulting in bone resorption, can also phagocytoze apoptotic bone cells [Boabaid et al., 2001; Cerri et al., 2003].

Myeloid derived phagocytes can be activated to play important roles during immune responses. Prostaglandin E_2 (PGE_2), a powerful short-lived lipid mediator, can shift the activated phagocytes to anti-inflammatory phenotypes, to cause general anti-inflammation and immune suppression. On the other hand, the recognition and removal of apoptotic cells by phagocytes is mediated by the exposure of phosphatidylserine (PS) on the cell membrane. In addition, increasing evidences has shown the PS-dependent phagocytosis of apoptotic cells to transform macrophages into an anti-inflammatory phenotype during an immune response [Fadok et al., 1998; Hoffmann et al., 2001; Huynh et al., 2002]. Moreover, liposomes containing PS (PS liposomes) can mimic the effects of apoptotic cells on the myeloid derived phagocytes including macrophages, microglia and dendritic cells to shift them from pro- to anti-inflammatory phenotype through the secretion of anti-inflammatory mediators including PGE_2 [Otsuka et al., 2004; Zhang et al., 2006; Shi et al., 2007].

This chapter first summarizes how PGE_2 modulates the differentiation and functions of myeloid derived phagocytes within immune responses, and then the potential therapeutic application of PS liposomes on inflammatory and immune diseases through feedback mechanisms involving PGE_2 will be discussed.

Effects of PGE_2 on Myloid-Derived Phagocytes: Macrophages

Macrophages modulate immune processes at sites of infection and inflammation through their cytokine production. PGE_2 plays an important role in limiting the inflammatory responses by regulating the cytokine synthesis by macrophages [Fennekohl et al., 2002; Rouzer et al., 2004; Chae et al., 2008].

PGE_2 is a negative regulator for pro-inflammatory cytokines in activated macrophages. Tumor necrosis factor-α (TNF-α) is a typical pro-inflammatory cytokine which promotes the immune processes in septic shock, rheumatoid arthritis, cancer, and diabetes. Endogenously produced PGE_2 acts in an autocrine fashion to limit TNF-α production by macrophages and/or peripheral blood monocytes in response to lipopolysaccharide (LPS) [Dieter et al., 1999;Fennekohl et al., 2002; Rouzer et al., 2004; Chae et al., 2008]. The inhibitory effect of PGE_2 on LPS-induced TNF-α production by macrophages is through increased level of intracellular cAMP, which selectively activates protein kinase A type I to inhibit TNF-α gene transcription [Stafford and Marnett, 2008]. In addition, PGE_2 attenuates the LPS-induced expression of chemokines including monocyte chemoattractant protein-1, interleukin (IL)-8, macrophage inflammatory protein-1α and -1β, and interferon (IFN)-inducible protein-10, and thereby inhibiting the IL-10 production [Takayama et al., 2002]. A selective EP4 receptor antagonist completely reversed PGE_2-mediated suppression of chemokine production [Takayama et al., 2002]. Moreover, EP4RAG, an EP4 receptor agonist, suppressed pro-inflammatory cytokines and chemokines by inhibiting the activation of nuclear factor (NF)-κB [Ogawa et al., 2009]. Therefore, the inhibitory effect of PGE_2 on the production of pro-inflammatory mediators by macrophages may be associated with the inhibition of NF-κB activity via an EP4 receptor [Takayama et al., 2002; Ogawa et al., 2009].

On the other hand, PGE_2 is a positive regulator of anti-inflammatory cytokines in the activated macrophages. IL-10, a typical anti-inflammatory cytokine, can attenuate immune processes. PGE_2 has been shown to enhance IL-10 production by cultured macrophages due to its cAMP-inducing capacity. *In vivo* experiments have suggested that increased PGE_2 production in a mycobacterium microti-infected mouse is closely linked with increased IL-10 production by peritoneal macrophages, because the addition of indomethacin, a preferential cyclooxygenase (COX)-1 inhibitor, causes a significant decrease in the IL-10 level [Mittal et al., 2001]. Mycobacterial products induce PGE_2-releasing macrophages in the mouse spleen. The isolated spleen cells from those infected mice express IL-10. The treatment of mice with NS-398, a selective COX-2 inhibitor, has been shown to result in the reduction of IL-10 production in recall responses, thus suggesting that the formation of splenic PGE_2-releasing macrophages is associated with the expression of COX-2 [*Shibata* et al., 2005]. The levels of LPS-stimulated production of PGE_2 by peritoneal macrophages are remarkably increased in pristane-induced lupus mice. Exogenous PGE_2 enhanced production of IL-10. Indomethacin greatly inhibits LPS-induced production of IL-10 by macrophages from pristane-induced lupus mice [Chae et al., 2008].

Effects of PGE_2 on Myloid-Derived Phagocytes: Microglia

Microglia are derived from myeloid origin and are responsible for the innate immune response in the brain [Perry, 1988]. Brain injuries or immunological stimuli, usually activate microglia to release a variety of soluble factors, which are pro-inflammatory in nature and potentially cytotoxic [Kreutzberg, 1996; Liu and Hong, 2003]. PGE_2, one of the major prostaglandins produced in the central nervous system (CNS), potently and selectively inhibits the TNF-α production but enhanced the IL-10 production in LPS-stimulated primary microglia [Aloisi et al., 1999; Petrova et al., 1999]. Agents that increase the cAMP levels mimic the action of PGE_2 on cytokine secretion, thus indicating the involvement of cAMP-coupled prostaglandin receptors. The local concentration ratio of PGE_2 can reduce the activity of microglia in the damaged CNS [Aloisi et al., 1999]. The inhibitory effect of PGE_2 on LPS-induced IL-18 production depends on the suppression of LPS-induced NF-κB activation [Suk et al., 2001]. Furthermore, PGE_2 also reduces the release of nitric oxide from microglia [Kim et al., 2002]. Moreover, PGE_2 can stimulate hippocampal granule cell neurogenesis [Uchida et al., 2002]. Therefore, these results indicate that exogenous PGE_2 protects cells from neuronal death through modulating the pro-and anti-inflammatory mediators secreted from microglia.

Effects of PGE_2 on Myloid-Derived Phagocytes: Dendritic Cells

Dendritic cells (DCs) play a major role in the induction of primary immune responses and tolerance [Banchereau and Steinman, 1998]. PGE_2 is produced in large amounts by activated DCs, and acts as a key modulator of their functions in an autocrine way [Harizi et al., 2002]. PGE_2 has been shown to inhibit the LPS stimulated TNF-α release from murine bone marrow-derived dendritic cells. In addition, the administration of PGE_2 in vivo reduced serum TNF-α and intracellular TNF-α in peritoneal isolated cells [Vassiliou et al., 2003]. Both IL-4 and granulocyte/macrophage colony-stimulating factor induce the differentiation of monocyte-derived DCs, which release PGE_2 upon stimulation with LPS. Inhibitors of p38 mitogen-activated protein kinase (MAPK) block LPS-induced COX-2 expression and PGE_2 release, thus indicating that LPS-induced PGE_2 synthesis by DCs depends on activated p38 MAPK-induced COX-2 expression [Norgauer et al., 2003]. On the other hand, exogenous PGE_2 enhances the IL-10 production from bone marrow-derived DCs (BM-DCs). IL-10 down-regulates production of both IL-6 and TNF-α by BM-DCs [Harizi and Norbert, 2004]. Therefore, PGE_2 plays a crucial role in determining the DC phenotype. EP2 and EP4 receptors are efficient for mediating PGE_2-induced modulation of DC functions, because the expression of EP2 and EP4 receptors is enhanced by LPS in a dose-dependent manner. NS-398 suppresses LPS-enhanced expression of EP2/EP4 receptors. In addition, butaprost, an EP2 receptor agonist, and PGE_1 alcohol, an EP2 and EP2/EP4 receptor agonist, enhance the IL-10 production by DCs [Harizi et al., 2003]. Furthermore, there is increasing evidence that PGE_2 derived from mesenchymal stem cells (MSCs) can inhibit myeloid dendritic cells maturation and function by selectively interfering with the generation of immature DCs. In the presence of MSCs, monocytes supplemented with granulocyte-macrophage colony-stimulating factor (GM-CSF) and IL-4 do not stimulate the maturation of DCs, while also failing to induce T-cell activation or proliferation due to the lack of IL-12 production. NS-398 restores DC differentiation and function. Indeed, direct addition of PGE_2 to cultures of monocytes blocks their differentiation toward DCs in a manner similar as in MSCs [Spaggiari et al., 2009].

Effects of PGE_2 on Myloid-Derived Phagocytes: Osteoclasts

Osteoclasts have been reported to be differentiated from the myeloid precursors [Miyamoto et al., 2001; Teitelbaum et al., 2003]. However, PGE_2 stimulates osteoclast differentiation in murine cultured cells at a relatively high concentration [Hurley et al., 1998; Okada et al., 2000], PGE_2 has been shown strongly inhibits human osteoclast formation [Take et al., 2005]. In this study, the $CD14^+$ human peripheral blood mononuclear cells expressed EP2 and EP4 receptors, PGE_2 inhibits receptor activated NF-□B ligand (RANKL)-induced human osteoclast formation, and H-89 blocks the inhibitory effect of PGE_2 on $CD14^+$ cell cultures. The results indicate that the inhibitory effect of PGE_2 on human osteoclast formation is mediated by EP2/EP4 receptors. More interestingly, the conditioned medium of $CD14^+$ cells pretreated with PGE_2 inhibits RANKL-induced osteoclast formation not only in human $CD14^+$ cell cultures, but also in mouse macrophage cultures. A low concentration of PGE_2 (500 pg/ml) significantly inhibits osteoclast differentiation in rat culture systems consistent with the previous data in rat cultures [Roux et al., 1997; Itonaga et al., 1999]. The inhibitory effect of PGE_2 on osteoclastogenesis is significantly antagonized by SQ22536, an adenylate cyclase inhibitor, thus suggesting the involvement of EP2 and/or EP4 receptors that are coupled to adenylate cyclase in the effect of PGE_2 [Tanaka et al., 2004; Take et al., 2005]. The receptor RANK/RANKL pathway plays essential roles in differentiation and the fusion of the precursors into mature osteoclasts [Kong et al., 1999; Li et al., 2000]. PGE_2 suppresses the RANKL expression in the bone marrow cells during the early stage of osteoclast differentiation in a rat culture system (unpublished data). PGE_2 may partially inhibit osteoclastogenesis in an autocrine manner because osteoclasts precursors have the receptors for PGE_2 [Kobayashi et al., 2005].

Therefore, endogenous PGE_2 inhibits the maturation and shifts the myeloid derived phagocytes to an anti-inflammatory phenotype by regulating the production of mediators in an autocrine manner, in turn to play a key role in anti-inflammation and immune suppression (Figure 1).

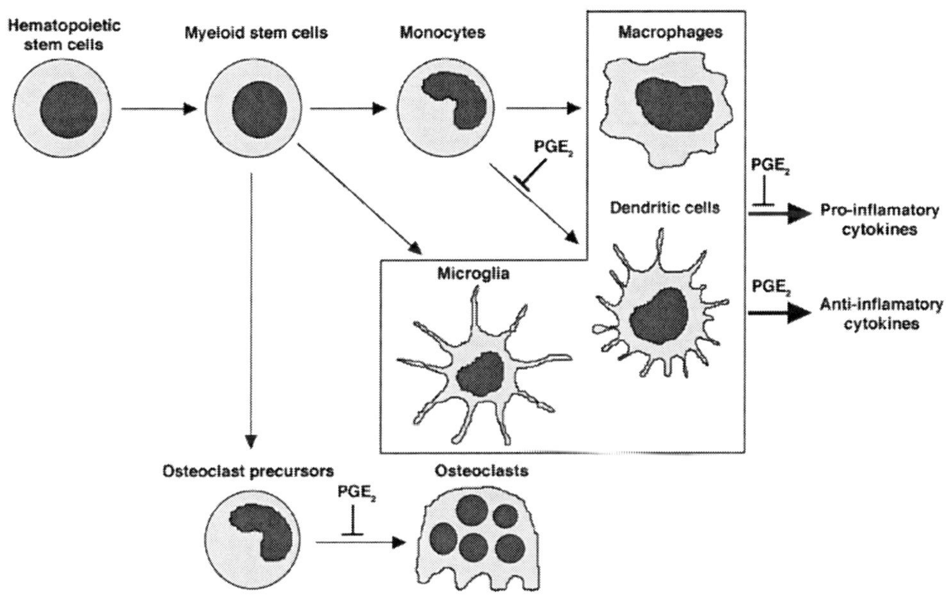

Figure1. A schematic illustration of the differentiation and functions of myeloid derived phagocytes regulated by PGE_2. PGE_2 inhibits the maturation of dendritic cells and the differentiation of osteoclasts. PGE_2 drives macrophages, microglia, and dendritic cells toward an anti-inflammatory phenotype during immune response.

Anti-Inflammatory Phenotypic Change of Myeloid – Derived Phagocytes Induced by PS Liposomes

Phagocytes which ingested apoptotic cells *in vitro* inhibit proinflammatory cytokine production through autocrine/paracrine mechanisms involving PGE_2. PS exposed on the apoptotic cell surface inhibits inflammatory responses, thus implying that PS may regulate the function of phagocytes. Apoptosis in vivo is followed almost inevitably by rapid uptake into adjacent phagocytic cells, which is a critical process in tissue remodeling, the regulation of the immune response, or the resolution of inflammation. In contrast, phagocytosis of apoptotic neutrophils actively inhibits the production of IL-1β, IL-8, granulocyte macrophage colony-stimulating factor, and TNF-α by human monocyte-derived macrophages. The addition of PGE_2 results in inhibition of LPS-stimulated cytokine production, thus suggesting that PGE_2 plays a role in phenotypic change of macrophages after phagocytosis of apoptotic cells [Fadok, 1998].

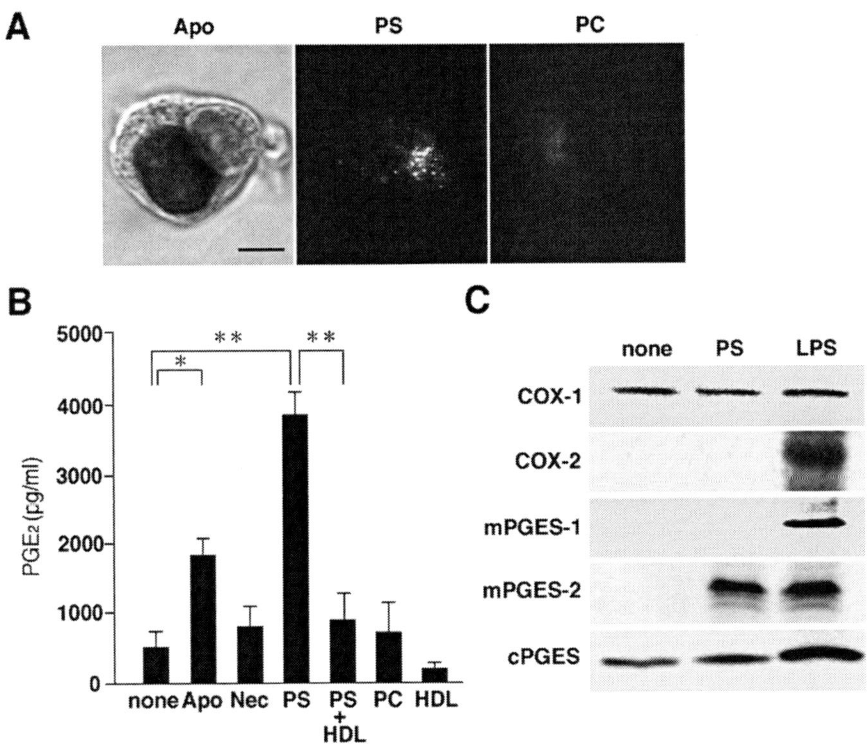

Figure 2. The production of PGE_2 after the phagocytosis of apoptotic neurons and PS liposomes by microglia. A: The phagocytosis of TUNEL-positive apoptotic neuron (Apo) and NBD-labeled PS liposomes (PS) but not PC liposomes (PC) by primary cultured microglia. Scale bar = 10 μm. B: The production of PGE_2 after phagocytosis of apoptotic neurons (Apo) and PS liposomes (PS) but not necrotic neurons (Nec) or PC liposomes (PC) by primary cultured microglia. The amount of PGE_2 in the culture medium was determined after 3 h treatment with Apo, Nec, PS (100 μM) or PC (100 μM) in the presence or absence of high density lipoprotein (HDL, 1 mg/ml). Apo and Nec were induced by treatment with sodium nitroproside (100 μM) and boiling, respectively. None: non-treatment. Each column and bar represents the mean ± S.E.M. of three experiments. Asterisks indicate a significanct difference between the values (*$P<0.05$; **$P<0.01$, Mann-Whitney U-test). C: Differential requirement of COX and PGES isoforms for PGE_2 production by microglia after treatment with PS liposomes and LPS. Microglial protein extracts after 3 h treatment with PS liposome (PS, 100 μM) or LPS (10 ng/ml) were subjected to immunoblot analyses. None: non-treatment.

PS liposomes can mimic the effects of apoptotic cells on macrophages, microglia and dendritic cells. A series of studies have shown that PS liposomes strongly reduce the release of pro-inflammatory molecules such as nitric oxide, IL-1β and TNF-α by LPS-activated microglia [De SR et al., 2002]. Microglia rapidly produce PGE_2 after treatment with PS liposomes. The specific ligand for class B scavenger receptor type I (SR-BI), high density lipoprotein, significantly suppresses both phagocytosis of PS liposomes and the subsequent PGE_2 production by microglia though the COX-1/microsonal prostaglandin E synthase-2 (mPGES-2) pathway (Figure 2) [Zhang et al., 2006].

PS liposomes inhibit the DC maturation to modulate adaptive immune responses. PS liposomes decreased the up-regulation of HLA-ABC, HLA-DR, CD80, CD86, CD40, and CD83, as well as the production of IL-12p70 by human DCs in response to LPS. DCs exposed to PS have diminished capacity to stimulate allogeneic T cell proliferation and to activate interferon (IFN)-γ-producing $CD4^+$ T cells [Chen et al., 2004]. An *in vivo* study has also demonstrated that PS liposomes reduce the expression of MHC II, CD80, CD86 and CD40 by DCs. Furthermore, PS liposomes inhibit IL-12 production but increase IL-10 production by DCs prepared from 1-chloro-2, 4-dinitrobenze (DNCB) challenged mice. PS liposomes also suppress the proliferation of $CD4^+$ T cell proliferation and IFN-γ production by DCs prepared from DNCB-challenged mice [Shi et al., 2007].

PS liposomes significantly inhibit the formation of mature osteoclasts but not that of osteoclast precursors. PS liposomes can be phagocytosed to enhance the secretion of PGE_2 quickly by the osteoclast precursors. PGE_2 itself can significantly inhibit osteoclast differentiation at a low concentration (500 pg/ml), and inhibitory effects of PS liposomes on the osteoclastogenesis is significantly antagonized by SQ22536, an adenylate cyclase inhibitor in our rat culture systems. The inhibitory effects of PS liposomes on the osteoclastogenesis may partly depend on PGE_2 secretion (unpublished data).

The PS liposome-induced modulation of myeloid derived phagocytes toward an anti-inflammatory phenotype may be achieved through the prompt PGE_2 production after phagocytosis by these cells. Therefore, PS liposomes can be potentially used as one type of pharmacological intervention for inflammatory and immune diseases.

PS Liposomes-Induced PGE_2 Production Pathway

The biosynthesis of PGE_2 occurs in a metabolic cascade starting from its precursor fatty acid, arachidonic acid (AA). AA is present in membrane phospholipids and released by the hydrolytic action of phospholipase A_2 (PLA_2). Released AA is metabolized to the unstable intermediate prostanoids by COX, also called PGH_2 synthase. COX has bifunctional catalytic properties. It oxygenates and cyclizes AA to form PGG_2 via its cyclooxygenase function, and this is followed by the reduction of PGG_2 to PGH_2 via its peroxidase function. PGH_2 is then converted to PGE_2 by the terminal synthases. PGE_2 synthesis is determined by the availability of the substrate AA and the COX activity. The rate-limiting enzyme COX become the focus of study since AA in total cellular fatty acid or in phospholipids fraction either remains unchanged or decreases with age. There are different isoenzymes of COX, COX-1, a constitutive form, and COX-2, an inducible form. COX-1 is constitutively expressed and is believed to be responsible for producing prostanoids to maintain physiological functions such as gastric protection and renal function. In contrast, COX-2 is regulated by growth factors, tumor promoters, cytokines, mitogens, glucocorticoids, and bacterial endotoxin, and is implicated in inflammatory responses and pathological changes in numerous disorders. PGES is not considered to be a rate-limiting factor in PGE_2 production, although recent publications suggests that it couples to COX and may participate in the regulation of PGE_2 synthesis. The activity and expression of mPGES-1 in peritoneal macrophages increases after pro-inflammatory stimuli. mPGES-1 is preferentially coupled with COX-2 in comparison to COX-1 [Murakami et al., 2000].

In contrast, microglia rapidly produce PGE_2 after treatment with PS liposomes. The PS liposome-induced PGE_2 production does not depend on COX-2, because neither COX-2 nor mPGES-1 is induced after treatment PS liposome. Instead, terminal PGESs, mPGES-2 and cytosolic PGES (cPGES), that are functionally coupled with COX-1 [Murakami et al., 2003] are markedly increased. Furthermore, PS liposome-induced PGE_2 production by microglia is significantly suppressed by indomethacin but not by NS-398. Therefore, COX-1 and its functionally coupled up-regulated mPGES-2 and cPGES, may play key roles in the PS liposome-induced PGE_2 production by microglia. Moreover, the phagocytosis of PS liposomes also promotes the PGE_2 production by transducing intracellular signals including extracellular signal-regulated kinase 1/2 but not p38 MAPK.

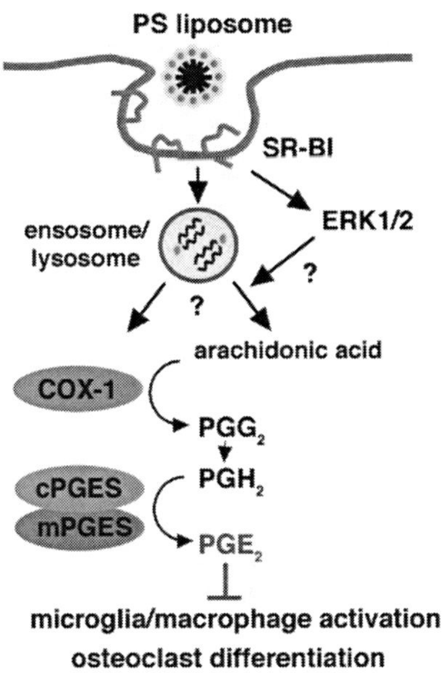

Figure 3. A schematic illustration of the pathway for the PGE_2 producton induced by PS liposomes and resultant possible potential anti-inflammatory interventions. SR-BI: class B scavenger receptor type I; ERK1/2: extracellular signal-regulated kinase 1/2; COX: cyclooxygenase; PG: prostaglandin; cPGES: cytosolic PGE synthase; mPGES: microsomal PGE synthase;

CONCLUSION

PS liposomes can inhibit the maturation and shift the myeloid derived phagocytes from a pro- to anti-inflammatory phenotype during an immune response. The effects of PS liposomes on myeloid derived phagocytes are mediated by PGE_2 produced through COX-1/mPGES-2 pathway in an autocrine manner (Figure 3). The clinical use of PS liposomes on the inflammatory and immune diseases without any deleterious side effects can therefore be expected because PS is a component of the cell membranes.

ACKNOWLEGMENTS

Studies conducted in our laboratories were supported by Grants-in-Aid for Scientific Research (No. 20390472 to HN, No. 20592174 to ZW) from the Ministry of Education, Svience and Culture, Japan.

REFERENCES

Aloisi, F., De Simone, R., Columba-Cabezas, S. & Levi, G. (1999). Opposite effects of interferon-g and prostaglandin E_2 on tumor necrosis factor and interleukin-10 production in microglia: a regulatory loop controlling microglia pro- and anti-inflammatory activities. *J. Neurosci. Res.*, *56*, 571-580.

Athanasou, N. A. & Quinn, J. (1990). Immunophenotypic differences between osteoclasts and macrophage polykaryons: immunohistological distinction and implications for osteoclast ontogeny and function. *J. Clin. Pathol.*, *43*, 997-1003.

Banchereau, J. & Steinman, R. M. (1998). Dendritic cells and the control of immunity. *Nature, 392*, 245-252.

Boabaid, F., Cerri, P. S. & Katchburian, E. (2001). Apoptotic bone cells may be engulfed by osteoclasts during alveolar bone resorption in young rats. *Tissue Cell*, *33*, 318-325.

Cerri, P. S., Boabaid, F. & Katchburian, E. (2003). Combined TUNEL and TRAPmethods suggest that apoptotic bone cells are inside vacuoles of alveolar bone osteoclasts in young rats. *J. Periodontal. Res.*, *38*, 223-226.

Chae, B. S., Shin, T. Y., Kim, D. K., Eun, J. S., Leem, J. Y. & Yang, J. H. (2008). Prostaglandin E_2-mediated dysregulation of proinflammatory cytokine production in pristane-induced lupus mice. *Arch. Pharm. Res.*, *31*, 503-510.

Chen, X., Doffek, K., Sugg, S. L. & Shilyansky, J. (2004). Phosphatidylserine regulates the maturation of human dendritic cells. *J Immunol.*, *173*, 2985-2994.

De, S. R., Ajmone-Cat, M. A., Nicolini, A. & Minghetti, L. (2002). Expression of phosphatidylserine receptor and down-regulation of pro-inflammatory molecule production by its natural ligand in rat microglial cultures. *J. Neuropathol. Exp. Neurol.*, *61*, 237-244.

Dieter, P., Hempel, U., Kamionka, S., Kolada, A., Malessa, B., Fitzke, E. & Tran-Thi, T. A. (1999). Prostaglandin E_2 affects differently the release of inflammatory mediators from resident macrophages by LPS and muramyl tripeptides. *Mediators Inflamm.*, *8*, 295-303.

Fadok, V. A., Bratton, D. L., Konowal, A., Freed, P. W., Westcott, J. Y. & Henson, P. M. (1998). Macrophages that have ingested apoptotic cells *in vitro* inhibit proinflammatory cytokine production through autocrine/paracrine mechanisms involving TGF-α, PGE_2, and PAF. *J. Clin. Invest.*, *101*, 890-898.

Fennekohl, A., Sugimoto, Y., Segi, E., Maruyama, T., Ichikawa, A. & Püschel, G. P. (2002). Contribution of the two Gs-coupled PGE_2-receptors EP2-receptor and EP4-receptor to the inhibition by PGE_2 of the LPS-induced TNF-α-formation in Kupffer cells from EP2-or EP4-receptor-deficient mice. Pivotal role for the EP4-receptor in wild type Kupffer cells. *J. Hepatol.*, *36*, 328-334.

Harizi, H., Grosset, C. & Gualde, N. (2003). Prostaglandin E_2 modulates dendritic cell function via EP2 and EP4 receptor subtypes. *J. Leukoc. Biol.*, *73*, 756-763.

Harizi, H., Juzan, M., Pitard, V., Moreau, J. F. & Gualde, N. (2002). Cyclooxygenase-2-issued prostaglandin E_2 enhances the production of endogenous IL-10, which down-regulates dendritic cell functions. *J. Immunol.*, *168*, 2255-2263.

Harizi, H. & Norbert, G. (2004). Inhibition of IL-6, TNF-□, and cyclooxygenase-2 protein expression by prostaglandin E_2-induced IL-10 in bone marrow-derived dendritic cells. *Cell. Immunol.*, *228*, 99-109.

Hurley, M. M., Lee, S. K., Raisz, L. G., Bernecker, P. J. & Lorenzo, J. (1998). Basic fibroblast growth factor induces osteoclast formation in murine bone marrow cultures. *Bone*, *22*, 309-316.

Hoffmann, P. R., deCathelineau, A. M., Ogden, C. A., Leverrier, Y., Bratton, D. L., Daleke, D. L., Ridley, A. J., Fadok, V. A. & Henson, P. M. (2001). Phosphatidylserine (PS) induces PS receptor-mediated macropinocytosis and promotes clearance of apoptotic cells. *J. Cell Biol.*, *155*, 649-659.

Huynh, M. L., Fadok, V. A. & Henson, P. M. (2002). Phosphatidylserine- dependent ingestion of apoptotic cells promotes TGF-□1 secretion and the resolution of inflammation. *J. Clin. Invest.*, *109*, 41-50.

Itonaga, I., Sabokbar, A., Neale, S. D. & Athanasou, N. A. (1999). 1, 25- Dihydroxyvitamin D_3 and prostaglandin E_2 act directly on circulating human osteoclast precursors. *Biochem. Biophys. Res. Commun.*, *264*, 590-595.

Iyoda, T., Shimoyama, S., Liu, K., Omatsu, Y., Akiyama, Y., Maeda, Y., Takahara, K., Steinman, R. M. & Inaba, K. (2002). The $CD8^+$ dendritic cell subset selectively endocytoses dying cells in culture and in vivo. *J. Exp Med.*, *195*, 1289-1302.

Kim, E. J., Kwon, K. J., Park, J. Y., Lee, S. H., Moon, C. H. & Baik, E. J. (2002). Neuroprotective effects of prostaglandin E_2 or cAMP against microglial and neuronal free radical mediated toxicity associated with inflammation. *J. Neurosci. Res.*, *70*, 97-107.

Kreutzberg, G. W. (1996). Microglia: a sensor for pathological events in the CNS. *Trends Neurosci.*, *19*, 312-318.

Kobayashi, Y., Take, I., Yamashita, T., Mizoguchi, T., Ninomiya, T., Hattori, T., Kurihara, S., Ozawa, H., Udagawa, N. & Takahashi, N. (2005). Prostaglandin E_2 receptors EP2 and EP4 are down-regulated during differentiation of mouse osteoclasts from their precursors. *J. Biol. Chem.*, *280*, 24035-24042.

Kong, Y. Y., Yoshida, H., Sarosi, I., Tan, H. L., Timms, E., Capparelli, C., Morony, S., Oliveira-dos-Santos, A. J., Van, G., Itie, A., Khoo, W., Wakeham, A., Dunstan, C. R.,

Lacey, D. L., Mak, T. W., Boyle, W. J. & Penninger, J. M. (1999). OPGL is a key regulator of osteoclastogenesis, lymphocyte development and lymph-node organogenesis. *Nature*, *397*, 315-323.

Li, J., Sarosi, I., Yan, X.Q., Morony, S., Capparelli, C., Tan, H. L., McCabe, S., Elliott, R., Scully, S., Van, G., Kaufman, S., Juan, S. C., Sun, Y., Tarpley, J., Martin, L., Christensen, K., McCabe, J., Kostenuik, P., Hsu, H., Fletcher, F., Dunstan, C. R., Lacey, D. L. & Boyle, W. J. (2000). RANK is the intrinsic hematopoietic cell surface receptor that controls osteoclastogenesis and regulation of bone mass and calcium metabolism. *Proc. Natl. Acad. Sci.*, *97*, 1566-1571.

Liu, B. & Hong, J. S. (2003). Role of microglia in inflammation-mediated neurodegenerative diseases: mechanisms and strategies for therapeutic intervention. *J. Pharmacol. Exp. Ther.*, *304*, 1-7.

Mittal, J., Dogra, N., Vohra, H. & Majumdar, S. (2001). Effects of prostaglandin E_2 and nitric oxide inhibitors on the expression of interleukin-10, interleukin-12 and MHC class-II molecules in Mycobacterium microti-infected and interferon-gamma-treated mouse peritoneal macrophages. *Folia Microbiol. (Praha)*, *46*, 259-264.

Miyamoto, T., Phneda, O., Arai, F., Iwamoto, K., Okada, S., Takagi, K., Anderson, D. & Suda, T. (2001). Bifurcation of osteoclasts and dendritic cells from common progenitors. *Blood*, *98*, 2544-2554.

Murakami, M., Nakatani, Y., Kuwata, H. & Kudo, I. (2000). Cellular components that functionally interact with signaling phospholipase A_2s. *Biochim. Biophys. Acta.*, *1488*, 159-166.

Murakami, M., Nakashima, K., Kamei, D., Masuda, S., Ishikawa, Y., Ishii, T., Ohmiya, Y., Watanabe, K. & Kudo, I. (2003). Cellular prostaglandin E_2 production by membrane-bound prostaglandin E synthase-2 via both cyclooxygenases-1 and -2. *J. Biol. Chem.*, *278*, 37937-37947.

Norgauer, J., Ibig, Y., Gmeiner, D., Herouy, Y. & Fiebich, B. L. (2003). Prostaglandin E_2 synthesis in human monocyte-derived dendritic cells. *Int. J. Mol. Med.*, *12*, 83-86.

Ogawa, M., Suzuki, J., Kosuge, H., Takayama, K., Nagai, R. & Isobe, M. (2009). The mechanism of anti-inflammatory effects of prostaglandin E_2 receptor 4 activation in murine cardiac transplantation. *Transplantation*, *87*, 1645-1653.

Okada, Y., Lorenzo, A., Freeman, A. M., Tomita, M., Morham, S. G., Raisz, L. G. & Pilbeam, C. C. (2000). Prostaglandin G/H synthase-2 is required for maximal formation of osteoclast-like cells in culture. *J. Clin. Invest.*, *105*, 823-832.

Otsuka, M., Tsuchiya, S. & Aramaki, Y. (2004). Involvement of ERK, a MAP kinase, in the production of TGF-β by macrophages treated with liposomes composed of phosphatidylserine. *Biochem. Biophys. Res. Commun.*, *324*, 1400-1405.

Perry, V. H. & Gordon, S. (1998). Macrophages and microglia in the nervous system. Trends Neurosci., *11,* 273-277.

Petrova, T. V., Akama, K. T. & Van Eldik, L. J. (1999). Selective modulation of BV-2 microglial activation by prostaglandin E_2. Differential effects on endotoxin-stimulated cytokine induction. *J. Biol. Chem.*, *274*, 28823-28827.

Roux, S., Pichaud, F., Qinn, J., Lalande, A., Morieux, C., Jullienne, A. & de Vernejoul, M. C. (1997). Effects of prostaglandins on human hematopoietic osteoclast precursors. *Endocrinology*, *138*, 1476-1482.

Rouzer, C. A., Kingsley, P. J., Wang, H., Zhang, H., Morrow, J. D., Dey, S. K. & Marnett, L. J. (2004). Cyclooxygenase-1-dependent prostaglandin synthesis modulates tumor necrosis factor-alpha secretion in lipopolysaccharide-challenged murine resident peritoneal macrophages. *J. Biol. Chem.*, *279*, 34256-34268.

Shi, D., Fu, M., Fan, P., Li, W., Chen, X., Li, C., Qi, X., Gao, T. & Liu, Y. (2000). Artificial phosphatidylserine liposome mimics apoptotic cells in inhibiting maturation and immunostimulatory function of murine myeloid dendritic cells in response to 1-chloro-2,4-dinitrobenze *in vitro*. *Arch. Dermatol. Res.*, *299*, 327-336.

Shibata, Y., Nishiyama, A., Ohata, H., Gabbard, J., Myrvik, Q. N. & Henriksen, R. A. (2005). Differential effects of IL-10 on prostaglandin H synthase-2 expression and prostaglandin E_2 biosynthesis between spleen and bone marrow macrophages. *J. Leukoc. Biol.*, *77*, 544-551.

Spaggiari, G. M., Abdelrazik, H., Becchetti, F. & Moretta, L. (2009). MSCs inhibit monocyte-derived DC maturation and function by selectively interfering with the generation of immature DCs: central role of MSC-derived prostaglandin E_2. Blood. *113*, 6576-83.

Stafford, J. B. & Marnett, L. J. (2008). Prostaglandin E_2 inhibits tumor necrosis factor-a RNA through PKA type I. *Biochem. Biophys. Res. Commun.*, *366*, 104-109.

Steinman, R. M., Turley, S., Mellman, I. & Inaba, K. (2000). The induction of tolerance by dendritic cells that have captured apoptotic cells. *J. Exp. Med.*, *191*, 495-502.

Suk, K., Yeou Kim, S. & Kim, H. (2001). Regulation of IL-18 production by IFN-γ and PGE□ in mouse microglial cells: involvement of NF-κB pathway in the regulatory processes. Immunol Lett., *77*, 79-85.

Takayama, K., García-Cardena, G., Sukhova, G.K., Comander, J., Gimbrone, M.A. Jr. & Libby, P. (2002). Prostaglandin E_2 suppresses chemokine production in human macrophages through the EP4 receptor. *J. Biol. Chem.*, *277*, 44147-44154.

Take, I., Kobayashi, Y., Yamamoto, Y., Tsuboi, H., Ochi, T., Uematsu, S., Okafuji, N., Kurihara, S., Udagawa, N. & Takahashi, N. (2005). Prostaglandin E_2 strongly inhibits human osteoclast formation. *Endocrinology*, *146*, 5204-5014.

Tanaka, M., Sakai, A., Uchida, S., Tanaka, S., Nagashima, M., Katayama, T., Yamaguchi, K. & Nakamura, T. (2004). Prostaglandin E_2 receptor (EP4) selective agonist (ONO-4819.CD) accelerates bone repair of femoral cortex after drill-hole injury associated with local upregulation of bone turnover in mature rats. *Bone*, *34*, 940-948.

Teitelbaum, S. L. & Ross, F. P. (2003). Genetic regulation of osteoclast development and function. *Nat. Rev. Genet.*, *4*, 638-649.

Uchida, K., Kumihashi, K., Kurosawa, S., Kobayashi, T., Itoi, K. & Machida, T., (2002). Stimulatory effects of prostaglandin E_2 on neurogenesis in the dentate gyrus of the adult rat. *Zoolog. Sci.*, *19*, 1211-1216.

Vassiliou, E., Jing, H. & Ganea, D. (2003). Prostaglandin E_2 inhibits TNF production in murine bone marrow-derived dendritic cells. *Cell. Immunol.*, *223*, 120-132.

Zhang, J., Fujii, S., Wu, Z., Hashioka, S., Tanaka, Y., Shiratsuchi, A., Nakanishi, Y. & Nakanishi, H. (2006). Involvement of COX-1 and up-regulated prostaglandin E synthases in phosphatidylserine liposome-induced prostaglandin E_2 production by microglia. *J. Neuroimmunol.*, *172*, 112-120.

In: Prostaglandins: Biochemistry, Functions...
Editors: Gillian M. Goodwin, pp.93-101 ISBN: 978-1-61668-272-9

Chapter 6

CROSS TALK BETWEEN THE RECEPTORS OF INSULIN AND PROSTACYCLIN ON PLATELETS MEMBRANE IN THE ENHANCEMENT OF ANTITHROMBOTIC EFFECT OF INSULIN

A. K. Sinha and M. Bhattacharyya
Sinha Institute of Medical Science & Technology,
288, Kendua main Road, Garia, Calcutta- 700084, India

ABSTRACT

Background

The aggregation of platelets by aggregating agents is essential for normal blood coagulation process. In contrast excessive platelet aggregation is reported to result in the coronary artery disease (CAD). The homeostasis is achieved through the inhibition of platelet aggregation by several humoral factors including prostacyclin (PGI_2) and insulin. Prostacyclin, a potent inhibitor of platelet aggregation, is however without thrombolytic effect. Insulin, also a potent inhibitor of platelet aggregation is known to mediate its effect through synthesis of nitric oxide (NO) which is a potent inhibitor of platelet aggregation and a thrombolytic agent. We have reported before that the treatment of platelets with PGI_2 increased the insulin receptor density on the platelet surface without receptor synthesis; due to the exposure of the "spare" insulin receptors on the membrane from the membrane bilayer. The effect of the increased insulin receptor numbers in platelets due to the "cross talk" with PGI_2 receptors on the antithrombotic effect of insulin was determined.

Methods

the binding of the ligands [^{125}I-insulin and ^{3}H PGE_1, a stable probe for PGI_2 receptor] was determined by Scatchard plot analysis of the equilibrium binding of the agonists using membrane filtration technique. Nitric oxide was determined by methemoglobin method. Thrombolysis *in vitro* was determined by clot lysis spectrophotometrically.

Results

The treatment of platelets in Tyrod's buffer with 200µunits of insulin in the presence or absence of 2nM PGE_1 for 2 h (equilibrium binding for insulin) resulted in the increase of insulin receptor numbers (n) from ≈120 sites/platelets in the control to 250 sites/platelets that increased the formation of NO from 1.2 nmol/mg/h in the control to 2.5nmol/mg/h in the presence of PGE_1 which resulted in the increased rate of clot lysis by ≈2 fold compared to control.

Conclusion

The systemic increase of PGI_2 might increase the insulin binding to platelets through the increase of insulin receptor numbers on the cell surface. The increased binding of insulin would not only inhibit platelet aggregation, but also enhance the thrombolytic effect of the hormone *in vivo*.

Platelet Aggregation and the Development of Coronary Artery Disease

Aggregation of human blood platelets by aggregating agents like ADP, l-epinephrine, collagen or thrombin is critically important in the life saving blood coagulation process (Furman MI et al., 1998 ; Massberg S et al., 2003). In contrast, excess platelet aggregation at the site of atherosclerotic plaque rupture on the coronary artery wall results in the development of coronary artery disease (CAD) in>90% of the cases due to the formation of a thrombus. The thrombus (actually a microaggergate of platelets embedded in fibrin mass) formed on the site of atherosclerotic plaque rupture on the wall of the artery serving the heart muscle (Falk E, 1992). The formation of the thrombus due to platelet aggregation would not only physically impede the normal blood circulation in the heart, the aggregation of platelets also resulted in the release of ADP from the platelets that led to the more platelet aggregation (Remijn JA et al., 2002). The ADP induced platelet aggregation produces several proaggregatory prostaglandins including prostaglandin G_2 which itself is a potent aggregator of platelets (Hamberg M et al., 1974). Although, as described above, there are several platelet aggregating agents, the aggregation of platelets by ADP is considered to be most important in the genesis of CAD at least in human (Mills DC, 1996). Development of the thrombus, however not only blocks the normal blood circulation in the heart, but it also interrupts the supply of oxygen, nutrients, minerals to the heart tissue, that in severe cases may result in the death of cardiac cells leading to the development of acute forms of the CAD like acute myocardial infarction (AMI) or unstable angina. And, as such, the formation of the thrombus on the arterial wall of the heart is considered to be the most important pathological event in the precipitation of CAD, which almost immediately results in the expression of the acute syndrome associated with the acute form of the condition (Falk E, 1992 ; Fuster V et al., 1996).

As described above, excess platelet aggregation at the site of the atherosclerotic plaque rupture precipitated CAD through thrombus formation. Conversely, the inhibition of platelet aggregation has been reported to prevent the occurrence of CAD, through the inhibition of thrombus formation (Fuster V et al., 1996). The chronic use of aspirin, a well-known inhibitor of platelet aggregation has been reported to decrease the risk of CAD through the inhibition of platelet aggregation *in vivo* (Antman EM et al.,1998).

Under physiologic conditions, chronic aggregation of platelets by the aggregating agents is counteracted by several inhibitory agents including prostacyclin (PGI_2) (Whittle BJ et al., 1978), interferon-α (IFN-α) (Bhattacharyya M et al., 2009), and insulin (Trovati M et al., 1997). While prostacyclin, an arachidonic acid metabolite of the cells (Marcus AJ et al., 1978) is a potent inhibitor of platelet aggregation through the stimulation of cyclic AMP level due to the activation of adenylate cyclase (Dutta-Roy AK et al., 1987), both insulin (Trovati M et al., 1997) and IFN-α (interferon-α), also potent inhibitors of platelet aggregation that inhibit platelet aggregation through the stimulation of nitric oxide synthesis (NO). Nitric oxide is not only a potent inhibitor of platelet aggregation (Chakraborty K et al., 2003), but also it can directly activate plasminogen to plasmin (Karmohapatra SK et al., 2007). Plasmin thus produced is a well known thrombolytic agent which also inhibits platelet aggregation through the dissolution of inter-platelet fibrinogen bridges that formed during the platelet aggregation without increasing platelet cyclic AMP or cyclic GMP levels (Karmohapatra SK et al., 2007 ; Bhattacharyya M et al., 2009). It is also reported that among the known inhibitors of platelet aggregation, NO alone is capable of disaggregating previously aggregated platelets affected by the NO induced formation of plasmin from plasminogen in the plasma. Both insulin and IFN-α are capable of activating plasminogen to plasmin that is capable of dissoluting previously clotted platelet-rich plasma produced by the addition of $CaCl_2$, that was accompanied by the appearance of fibrin degradation products in the assay mixture (Bhattacharyya M et al., 2009; Karmohapatra SK et al., 2007). These reports indicated that the dissolution of the clotted platelet-rich plasma by the above proteins was mediated by fibrinolysis catalyzed by plasmin formed due to the activation of plasminogen to plasmin (Karmohapatra SK et al., 2007). It has been reported that the injection of insulin in the circulation of mice pretreated with ADP to develop coronary artery thrombosis not only dissolute the coronary thrombus in the animal model that could be visualized by light microscopy, but the injection of the hormone in these animals prevented death due to the ADP induced thrombus formation in the coronary artery of the animals (Sinha AK et al., 1999).

To determine the efficacy of insulin injection to prevent death in humans suffering from AMI or UA was carried in clinical trials where insulin was injected (6units/10kg weight /6 h) in these patients (n=138). It was found that none of these victims died, and none of the patients needed opioids to relieve the acute chest pain. Subsequent recuperation was uneventful (Kahn NN et al., 1992).

In contrast to the insulin, the use of PGI_2 one of the most potent inhibitors of platelet aggregation, and a much studied molecule, as a potential "cure" for CAD, has been turned out to be disappointing (Mueller HS et al., 1985; Mehta P et al., 1980).

The failure of PGI_2 as a curative therapeutic agent could be related to the fact that although the inhibition of platelet aggregation would result in the inhibition of thrombus formation at the site of the atherosclerotic plaque rupture, the dissolution of the formed thrombus, which in the first place, causes CAD is essential for the functional recovery from the condition. Despite the fact that PGI_2 is a potent inhibitor of platelet aggregation, it has no thrombolytic activity either *in vitro* or *in vivo* (Sinha A.K et al., 1999) may consequently fail to "cure" the established CAD.

In the above context, the role of PGI_2 as a systemic antithrombotic agent for the resolution of acute ischemic heart disease *in vivo* remains disappointing due to its lack of thrombolytic properties. However, this conclusion may be not be entirely correct in the backdrop of "cross-talk" between the receptors of insulin and PGI_2 on the platelet surface in relation to the antithrombotic effect of both the hormone and the prostanoid *in vivo*.

"Cross Talk" between the Receptors of Insulin and the Prostaglandin

In general sense the transduction of a particular hormone effect accepted to be the consequence of the interaction of the hormone (agonist) to its receptors on the cell surface. The interaction of the hormone with its receptor is related in the activation of the receptor. The activated receptors in turn transduce the hormonal message across the cell membrane, and the hormone effect is expressed. As such, the interaction between the agonist and the receptor is very specific and important in the expression of the hormone effect. Since the binding of the agonist to its receptor is ultimately resulted in the hormone effect, the availability of the receptor for its agonist is critically important in the hormone effect. Although it has been reported that in the cases of both PGI_2 and insulin that have their own specific receptor sites on the membrane surface, it has also been reported that the interaction of one of the agonists to its own receptors on the cell membrane resulted in the up regulation of the receptor numbers of the other agonists. In other words, binding of one of the agonists to its own receptors resulted in the increase of receptor numbers for the other agonist on the cell surface in the case of both platelets and erythrocytes (Kahn NN et al., 1990 ; Ray TK et al., 1986). The increased availability of the receptor numbers of the agonist consequently resulted in the increased expression of the agonist (hormone) effect on platelet or erythrocyte surface (Kahn NN et al., 1990 ; Ray TK et al., 1986). This increase of available receptors on the cell surface in the case of platelets or erythrocytes was not related to the synthesis of new protein receptor macromolecules because neither human erythrocytes nor platelets are capable of synthesizing protein. It has been postulated that this modulation of receptor numbers on the cell surface was the consequence of exposure of "spare" receptors from the membrane bilayer to the cell surface (Kahn NN et al., 1990). This "cross talk" between the receptors of insulin and PGI_2 has been reported to have important consequences on the antithrombotic effect of $PGI_{2,}$ and insulin through the upmodulation of the receptor numbers on the platelet surface that could have important effect both in the inhibition of platelet aggregation leading to the inhibition of thrombus formation and to the thrombolysis in CAD in the system.

The Upregulation of PGI_2 Receptor Numbers on the Platelet Surface by Insulin and the Consequent Effect on the Inhibition of Platelet Aggregation

Incubation of the isolated platelet membrane preparation with ^3H prostaglandin E_1, [a stable probe for PGI_2 receptor (Kahn NN et al., 1990) i.e., PGE_1 binds to the same PGI_2 receptors on the cell surface (Dutta-Roy AK et al., 1987)] indicates that PGI_2 binds to platelet surface on one high affinity (kd_1) low capacity (n_1) and one low affinity (kd_2) and high capacity (n_2) receptors populations. The membrane preparation was incubated with insulin (0.7nM) for 2.5h (to allow the binding of insulin to its receptor to reach equilibrium) before the incubation of the membrane preparation with ^3H PG E_1 for 20 mins (both incubation at 23°C) to allow the binding of the probe to its receptor to attain the equilibrium. When the binding characteristics of the prostaglandin binding was analyzed by scatchard plot (Kahn NN et al., 1990) it was found that both high affinity–low capacity and low affinity-high capacity prostaglandin receptor population increased by nearly 2 fold after the platelet membrane were incubated with insulin prior to the addition of $^3HPGE_1$ to the incubation mixture. It was found that the n_1 (high affinity –low capacity) of PGI_2 receptor population which was 1.6 pmol $^3HPGE_1$/mg membrane protein before the membrane preparation was incubated with insulin increased to 2.5 pmol PGE_1/mg membrane protein after the membrane was treated with insulin. Similarly the number of low affinity high capacity PGI_2 receptor population (i.e. n_2) increased from the basal 15.1 pmol/mg protein to 33.30 pmol/mg protein after the membrane preparation was treated with insulin. The incubation of the membrane preparation with insulin affected neither the kd_1 (9.2 nM) nor kd_2 (1.1μM). Use of platelet rich plasma instead of the isolated platelet membrane in the binding assay of $^3HPGE_1$ incubated previously with the hormone resulted in the similar increase of both n_1 and n_2 without changing the affinities (kd_1 and kd_2) of the binding.

The increased prostaglandin binding to platelets in platelet rich plasma preincubated with insulin as described above also resulted in the increased inhibition of ADP induced platelet aggregation by 2 fold compared to control. It was found that the minimal inhibitory concentration of PGE_1 needed for the maximal inhibition of platelet aggregation was 36.4±15.0 nM PGE_1 in the case of platelet rich plasma in control experiments. The minimum inhibitory dose of PGE_1 decreased to 15.7±10.3 nM in the case of the same platelet rich plasma which was preincubated with insulin before the addition of PGE_1 to inhibit platelet aggregation.

Effect of Prostacyclin in the Insulin Induced Thrombolysis

As discussed above PGI_2, although is a potent inhibitor of platelet aggregation (Whittle BJ et al., 1978), it has no thrombolytic activity, and, as such, it is not totally unexpected that clinical trials of PGI_2 in acute coronary artery disease did not measure up to the high expectation of the protagonists (Mueller HS et al., 1985 ; Mehta P et al., 1980). On the other hand it has been reported that insulin and aspirin, unlike PGI_2, were capable of thrombolysis both *in vitro* and *in vivo* through the stimulation of NO synthesis (Chakraborty K et al., 2004 ; Karmohapatra SK et al., 2007). Indeed for its multifaceted antithrombotic effects, insulin has been reported to be an antithrombotic humoral factor (Chakraborty K et al., 2004). All effects of insulin are however mediated through the binding of the hormone to its receptor on various cell surfaces (Czech M.P, 1977).

As described above the binding of insulin to the platelets and other cell increases the binding of PGI_2 and PGE_1 to different cells (Chakraborty S et al., 2004 ; Kahn NN et al., 1990 ; Ray TK. et al., 1986), the binding of PGE_1 (as a receptor probe for PGI_2) conversely increased the binding of insulin to different cells. Scatchared analysis of the equilibrium binding of insulin to its receptors to the platelets preparation that was incubated with 2.0 nM PGI_2 demonstrated that the physiologically important high-affinity low capacity insulin receptor numbers (n_1) on the platelet surface increased from the basal 120±20 sites/platelet to 250±28 sites/platelet without any change of the receptor affinity. The increase of insulin receptor numbers due to the incubation of the platelets with PGI_2 resulted in the increased production of NO in platelets. It was found that in the presence of insulin (200 μUnits/ml) the basal production of NO (1.2l±0.11 nmol NO/10^6platelets/h) increased to 2.6±0.15 nmol NO/10^6 latelets/h. The increased insulin production resulted in the clot lysis [platelet rich plasma clotted by the addition of $CaCl_2$ (Haskel EJ et al., 1993)] by≈ 2 fold through the activation of plasminogen to plasmin (Sinha AK et al., 1999).

Cross-Talk between Prostacyclin and Insulin Receptors on Erythrocytes Membrane

The cross-talk between the receptors of PGI_2 and insulin was however not a platelet specific phenomenon. Similar cross-talk between the receptors of the hormone and the prostaglandin in the case of human erythrocytes membrane has also been reported (Ray TK et al., 1986). It has been reported that the incubation of isolated erythrocyte membrane with PGI_2 (2.0 nM) resulted in the increase of high affinity insulin receptor numbers as in the case of platelets without affecting the affinity of the receptors. The increased insulin binding has been reported to decrease microviscosity of erythrocytes through the increased availability of the receptor on the erythrocytes membrane (Ray TK et al., 1986).

The effect of PGI_2 on the erythrocytes was highly specific. It has been reported that the binding to PGI_2 by other prostaglandins except PGE_1 does not compete with the PGI_2 receptor binding on the erythrocyte membrane (Ray TK et al., 1986). The increased insulin binding on the erythrocyte membrane in the presence of PGI_2, which itself has no effect on microviscosity, resulted in the decreased microviscosity of the erythrocyte membrane. Since deformability of erythrocytes which is inversely related to the microviscosity, the erythrocyte deformability is an important physiologic phenomenon for rapid and efficient diffusion of O_2 in the microvasculature. The heterologous upregulation of insulin receptor numbers achieved through the binding of PGI_2 to its own receptor indicated that the cross-talk between the receptors of hormone and PGI_2 in erythrocytes could be of physiologic significance.

Discussion

Hormone specific receptor interaction is generally involved in the expression of the hormonal effect. The post receptor interaction initiated through the activation of the hormone receptor is usually thought to be autonomous at least in the case of a specific hormonal interaction. This specificity is sometimes compared to that of enzyme-substrate specificity. The cross-talk between the receptors of different agonist contradicts the proposition that the expression of the hormonal effect *in vivo* is controlled by the hormone itself. As presented above the final outcome of the effects of one hormone is actually significantly controlled by the presence of another agonist that influences the expression of the effect of the hormone in question through the activation of a different receptor on the cell surface. In other words it is not the receptor occupancy alone that controls hormonal action *in vivo* in all cases, but different receptor occupancy by an independent agonist may play an important role in the process. Furthermore cross-talk between the hormone is not necessarily due to the synthesis of new receptors by target cells in question, but the cross-talk could take place in the absence of synthesis of new receptors macromolecule (Ray TK et al., 1986 ; Kahn NN et al., 1990).

In the case of insulin and PGI_2 receptors on platelets, the upregulation of PGI_2 receptors by insulin would result in the amplified inhibition of platelet aggregation that may prevent the development of coronary thrombosis on the vessel wall. On the other hand upregulation of the hormone receptor number by the prostaglandin on the platelet surface may actually have an important contribution in the thrombolytic process through the hormone induced NO synthesis *in vivo* through the generation of plasmin from plasma plasminogen (Karmohapatra SK et al., 2007). The lack of "curative" effect of PGI_2 may not actually represent the *in vivo* effect of PGI_2 in the thrombolysis in the presence of insulin in the system. However it should be mentioned here that patients who had no history of diabetes mellitus with acute ischemic heart disease are hyperglycemic due to lack of insulin synthesis in the system mimicking type-I diabetes mellitus. In this context it could perhaps be speculated that the failure of PGI_2 as a curative agent in acute CAD was at least in part due to severe deficiency of insulin in these patients.

Furthermore, as the binding of insulin to its receptors plays the essential role in the control of hyperglycemia for its antidiabetic effect, and since PGI_2 binding increases the hormone binding to its receptors the presence of the PGI_2 in the system will increase the binding of insulin for its antidiabetic effect.This cross talk *in vivo* may also amplify the antidiabetic effect of insulin essential for carbohydrate metabolism..

The cross-talk between insulin and PGI_2 is not only reported to be involved in the prevention and resolution of coronary thrombosis as described above, but the upregulation of insulin receptors numbers on the neutrophils in the presence of PGI_2 or PGE_1 reported to result in the synthesis of maspin, an antibreast cancer protein through the stimulated synthesis of NO in these cells (Girish GV et al., 2006).

In conclusion it could be said that the crosstalk between receptors of insulin and prostacyclin on different cells including platelets, neutrophils and erythrocytes regulates the effects of these agonists in wide variety of conditions including the resolution of coronary thrombosis, control of diabetes, the prevention of breast cancer as well as the perfusion of O_2 in the microcirculation.

References

Antman, EM; Braunwald, E. Acute myocardial infarction. In: Tinsley Randolph Harrison, S; Anthony; Fauci, Eugene Braunwald, J; Kurt, Isselbacher, D; Jean, Wilson, B; Joseph, Martin, L; Dennis, Kasper, L; Stephen, Hauser, L. Dan, Longo (Eds.), *Harrison's Principle of Internal Medicine*. McGraw-Hill Inc. New York, 1998, 1352-1365.

Bhattacharyya, M; Karmohapatra, SK; Bhattacharya, G; Bhattacharya, R; Sinha, AK. The role of leucocytes in the acetyl salicylic acid (aspirin) induced nitric oxide synthesis in the production of interferon-α, a potent inhibitor of platelet aggregation and a thrombolytic agent. *Journal throm and Thrombolysis.*, 2009, 28(2), 173-183.

Chakraborty, K; Khan, GA; Banerjee, P; Ray, U; Sinha, AK. Inhibition of human blood platelet aggregation and the stimulation of nitric oxide synthesis by aspirin. *Platelets.*, 2003, 14, 421-427.

Chakraborty, K; Sinha, AK. The role of insulin as an antithrombotic humoral factor. *Bioessays.*, 2004, Jan, 26(1), 91-8.

Chakraborty, S; Girish, GV; Sinha, AK. Impaired binding of insulin to erythrocyte membrane receptor and the activation of nitric oxide synthase by the hormone in human breast cancer. *J Cancer Res Clin Oncol.*, 2004 May, 130(5), 294-300. Epub, 2004 Feb 19.

Chakraborty, S; Girish, GV; Sinha, AK. Impaired binding of insulin to erythrocyte membrane receptor and the activation of nitric oxide synthase by the hormone in human breast cancer. *J Cancer Res Clin Oncol.*, 2004 May, 130(5), 294-300.

Czech, MP. Molecular basis of insulin action. *Annu Rev Biochem.*, 1977, 46, 359-84.

Dutta-Roy, AK; Sinha, AK. Purification and properties of prostaglandin E1/prostacyclin receptor of human blood platelets. *J Biol Chem.*, 1987, Sep, *15*, 262(26), 12685-91.

Falk, E. Dynamics in thrombus formation. *Ann NY Acad Sci.*, 1992, 667, 204-23.

Furman, MI; Benoit, SE; Barnard, MR; Valeri, CR; Borbone, ML; Becker, RC; Hechtman, HB; Michelson, AD. Increased platelet reactivity and circulating monocyte-platelet aggregates in patients with stable coronary artery disease. *J Am Coll Cardiol.*, 1998, 1, 352-358.

Fuster, V; Badimon, J; Chesebro, JH; Fallon, JT. Plaque rupture, Thrombosis, and therapeutic implications. *Haemostasis.*, 1996, 26, 269-284.

Girish, GV; Bhattacharya, G; Sinha, AK. The role of insulin dependent NO synthesis in the impaired production of maspin in human breast cancer. *J Cancer Res Clin Oncol.*, 2006, Jun, 132(6), 389-98.

Hamberg, M; Svensson, J; Wakabayashi, T; Sammuelsson, B. Isolation and structure of two prostaglandin endoperoxides that cause platelet aggregation. *Proc Natl Acad Sci*, U S A., 1974, 71(2), 345-9.

Haskel, EJ; Eisenberg, PR; Abendschein, DR. Inhibition of concomitant thrombin-mediated fibrin formation enhances clot lysis in whole blood. *Blood Coagul Fibrinolysis.*, 1993, Feb, 4(1), 7-13.

Kahn, NN; Bauman, WA; Hatcher, VB; Sinha, AK. Inhibition of platelet aggregation and the stimulation of prostacyclin synthesis by insulin in humans. *Am J Physiol.*, 1993, Dec, 265(6 Pt 2), H2160-7.

Kahn, NN; Najeeb, MA; Ishaq, M; Rahim, A; Sinha, AK. Normalization of impaired response of platelets to prostaglandin E1/I2 and synthesis of prostacyclin by insulin in unstable angina pectoris and in acute myocardial infarction. *Am J Cardiol.*, 1992, Sep 1, 70(6), 582-6.

Mueller, HS; Rao, PS; Greenberg, MA; Buttrick, PM; Sussman, II; Levite, HA; Grose, RM; Perez-Davila, V; Strain, JE; Spaet, TH. Systemic and transcardiac platelet activity in acute myocardial infarction in man: resistance to prostacyclin. *Circulation.*, 1985 Dec, 72(6), 1336-45.

Kahn, NN; Sinha, AK. Stimulation of prostaglandin E1 binding to human blood platelet membrane by insulin and the activation of adenylate cyclase. *J Biol Chem.*, 1990, Mar 25, 265(9), 4976-81.

Karmohapatra, SK; Chakraborty, K; Khann, NN; Sinha, AK. The role of nitric oxide in aspirin induced thrombolysis in vitro and the purification of aspirin activated nitric oxide synthesis from human blood platelets. *Am J Hem.*, 2007, 2, 986-995.

Karmohapatra, SK; Kahn, NN; Sinha, AK. The thrombolytic effect of aspirin in animal model. *J Thromb Thrombolysis.*, 2007, 24(2), 123-9.

Marcus, AJ; Weksler, BB; Jaffe, EA. Enzymatic conversion of prostaglandin endoperoxide H2 and arachidonic acid to prostacyclin by cultured human endothelial cells. *J Biol Chem.*, 1978, Oct 25, 253(20), 7138-41.

Massberg, S; Schulz, C; Gawaz, M. Role of platelets in the pathophysiology of acute coronary syndrome. *Semin Vasc Med.*, 2003, (2), 147-162.

Mehta, P; Mehta, J. Platelet function studies in coronary heart disease: VIII. Decreased platelet sensitivity to prostacyclin in patients with myocardial ischaemia. *Thromb Res.*, 1980 Apr 1-15, 18(1-2), 273-7.

Mills, DC. ADP receptors on platelets. *Thromb Haemost*, 1996, 6(6), 835-856.

Ray, TK; Dutta-Roy, AK; Sinha, AK. Regulation of insulin receptor activity of human erythrocyte membrane by prostaglandin E1. *Biochim Biophys Acta.*, 1986, Apr, 25, 856(3), 421-7.

Remijn, JA; Wu, YP; Jeninga, EH; IJsseldijk, MJ; van Willigen, G; de Groot, PG; Sixma, JJ; Nurden, AT; Nurden, P. Role of ADP receptor P2Y(12) in platelet adhesion and thrombus formation in flowing blood. *Arterioscler Thromb Vasc Biol., 2002, 22(4), 686-91.*

Sinha, AK; Bhattacharya, S; Acharya, K; Mazumder, S. Stimulation of nitric oxide synthesis and protective role of insulin in acute thrombosis in vivo. *Life Sci.*, 1999, 65(25), 2687-96.

Tateson, JE; Moncada, S; Vane, JR. Effects of prostacyclin (PGX) on cyclic AMP concentrations in human platelets. *Prostaglandins.*, 1977 Mar, 13(3), 389-97.

Trovati, M; Anfossi, G; Massucco, P; Mattiello, L; Costamagna, C; Piretto, V; Mularoni, E; Cavalot, F; Bosia, A; Ghigo, D. Insulin stimulates nitric oxide synthesis in human platelets and, through nitric oxide, increases platelet concentrations of both guanosine-3', 5'-cyclic monophosphate and adenosine-3', 5'-cyclic monophosphate. *Diabetes.*, 1997May, 46(5), 742-9.

Whittle, BJ; Moncada, S; Vane, JR. Comparison of the effects of prostacyclin (PGI2), prostaglandin E_1 and D_2 on platelet aggregation in different species. *Prostaglandins.*, 1978 Sep, 16(3), 373-88.

In: Prostaglandins: Biochemistry, Functions…
Editors: Gillian M. Goodwin, pp.103-111 ISBN: 978-1-61668-272-9

Chapter 7

IS MPGES-1 AN IDEAL TARGET FOR COUNTERACTING INFLAMMATION AND ESCAPING CARDIOVASCULAR SIDE EFFECTS ASSOCIATED WITH COX-2 INHIBITORS? A BRIEF OVERVIEW

Mohit Kapoor**, *Jean-Pierre Pelletier* and *Johanne Martel-Pelletier

Osteoarthritis Research Unit, University of Montreal Hospital Research Centre (CRCHUM), Notre-Dame Hospital, Montreal, Quebec, Canada

ABSTRACT

Non-steroidal anti-inflammatory drugs (NSAIDs) are extensively used for the management of inflammation, pain, swelling and stiffness associated with diseases such as rheumatoid arthritis, osteoarthritis, inflammatory arthropathies, post-operative pains, etc. NSAIDs act via inhibition of cyclooxygenases (COX) resulting in subsequent inhibition of prostaglandins (PGs) in the arachidonic acid (AA) metabolic pathway. Traditional non-selective NSAIDs including indomethacin, ibuprofen and naproxen, though effective in the management of pain and inflammation, were associated with a number of gastrointestinal (GI) and renal side effects because of their ability to inhibit activities of both isoforms of cyclooxygenase, (COX)-1 and -2, thus resulting in inhibition of PGs and thromboxane (TXs) production. Since COX-1 regulates GI, renal and vascular functions, selective inhibition of inducible COX-2 was thought to be a better alternative and as a result selective COX-2 inhibitors (coxibs) were developed. Coxibs including celecoxib, rofecoxib and valdecoxib were claimed to produce anti-inflammatory effects with low GI side effects. Although coxibs were associated with reduced incidence of serious GI side effects, these drugs demonstrated an increased risk for cardiovascular events associated with the inhibition of COX-2. For this reason, rofecoxib in late 2004 and valdecoxib in 2005 were withdrawn from the market. This withdrawal has raised several serious questions regarding the balance of safety and

* Corresponding author: Email: mohit.kapoor.chum@ssss.gouv.qc.ca

efficacy of this class of drugs and, as a result, alternative targets in the AA metabolic pathway are being identified to overcome the side effects associated with coxibs. Since the discovery of microsomal prostaglandin E synthase-1 (a PGE_2 inducible enzyme downstream of COX), several reports have suggested the possibility of mPGES-1 as a future therapeutic target in counteracting inflammation during inflammatory diseases. This review will discuss briefly the prospect of mPGES-1 inhibition as an alternative therapeutic strategy for counteracting inflammation. The question that we will put in perspective is whether its pharmacological inhibition will bypass the cardiovascular side effects associated with COX-2 inhibitors?

INTRODUCTION

In 1971, a breakthrough discovery by Nobel Laureate Sir John Robert Vane demonstrated that the mechanism of action of aspirin and other non-steroidal anti-inflammatory drugs (NSAIDs) was via inhibition of prostaglandin (PG) production[1]. Subsequently, in the 1990s two cyclooxygenase (COX) enzyme biosystems were discovered, which were involved in metabolizing arachidonic acid (AA) into physiologically active metabolites including PGs and thromboxanes (TXs)[2-4]. COX-1 is constitutively expressed in a variety of cells and tissues and regulates the production of PGs and TXs involved in the regulation of gastrointestinal (GI), renal and vascular homeostasis[5]. COX-2, on the other hand, is induced in inflammatory cells and tissues in response to a variety of pro-inflammatory stimuli and regulates the production of PGs, especially PGE_2 and PGI_2 (prostacyclin) that are involved in inflammation, pain and fever[6-9]. PGE_2 is the principal PG product during inflammatory diseases such as rheumatoid arthritis (RA)[8, 10-11] and osteoarthritis (OA)[6, 12] etc.

Traditional non-selective NSAIDs such as indomethacin, ibuprofen and naproxen inhibit PGE_2 production via inhibition of both COX-1 and COX-2, thus resulting in an anti-inflammatory effect. However, these drugs are associated with moderate to severe GI and renal side effects due to the inhibition of COX-1. Selective COX-2 inhibitors (coxibs) including celecoxib, rofecoxib and valdecoxib were subsequently generated to overcome the GI and renal side effects associated with non-selective NSAIDs[13]. The VIGOR (Vioxx GI Outcomes Research) study, which compared the efficacy and adverse effect profiles of rofecoxib and naproxen, showed that treatment with rofecoxib resulted in significantly less clinically important upper gastrointestinal events than treatment with naproxen in patients with RA[14]. Although rofecoxib exhibited a superior GI adverse effect profile over conventional NSAIDs, its long-term use was associated with increased risk of myocardial infarction and thrombosis[15]. For this reason, rofecoxib was withdrawn worldwide in 2004. In 2005, valdecoxib was also withdrawn by the United States Food and Drug Administration (FDA) with concerns over possible increased risk of heart attack and stroke. Celecoxib is currently under FDA alert [16].

Intense investigation is underway as to how the long-term use of coxibs results in serious cardiovascular side effects. One of the possible primary mechanisms is the ability of COX-2 inhibitors to not only inhibit PGE_2 production but to also cause imbalance in the production of prostaglandin I_2 (PGI_2) and TXA_2. Cardiovascular side effects of COX-2 inhibition may in part be caused by the loss of anti-thrombotic PGI_2 derived from endothelial COX-2, which plays a key role in the regulation of thrombogenesis[17]. Within the AA metabolic pathway,

COX-2 is placed relatively higher in the pathway resulting in inhibition of all PGs. Therefore, targeting specific inhibition of the major pro-inflammatory PG, PGE_2, while sparing PGI_2 and TXA_2 imbalance, seems to be an interesting alternative to achieve a specific anti-inflammatory effect and escape loss of vascular function and homeostasis.

MPGES-1

The conversion of COX-derived PGH_2 to PGs was believed to be a non-enzymatic process. However, in recent years, various terminal synthases that act in the conversion of PGH_2 to each active prostanoid downstream of COX have been cloned and characterized. In 1999, the protein directly involved in PGE_2 production within the AA metabolic pathway was identified and named microsomal PGE synthase-1 (mPGES-1)[18]. This elegant work showed that human microsomal glutathione transferase1-like-1, a member of the membrane-associated proteins in eicosanoid and glutathione metabolism (MAPEG) superfamily expressed in Escherichia coli, exhibited PGE_2 synthase activity. mPGES-1 was shown to be an inducible enzyme that was induced in response to pro-inflammatory cytokines. In addition to mPGES-1, two other PGE_2 synthases (mPGES-2 and cytosolic PGES [cPGES][18-20]) were identified and cloned.. cPGES, initially identified as heat shock protein 90 (Hsp90)-associated protein p23, is constitutive and is functionally coupled with COX-1 in the cytosol under basal conditions in various cells and tissues[20]. mPGES-2 is also constitutively expressed in a variety of cells and tissues. However, it is functionally coupled with both COX-1 and COX-2[21], suggesting its role in the production of PGE_2 not only for homeostasis but also during disease processes.

Role of mPGES-1 in Inflammation

mPGES-1 acts downstream of COX enzymes and specifically converts PGH_2 to PGE_2 under basal as well as inflammatory conditions[22]. *In vitro* studies first demonstrated that mPGES-1 is induced in response to different pro-inflammatory stimuli including lipopolysaccharide (LPS), interleukin-1beta (IL-1β) and tumor necrosis factor-alpha (TNF-α) in a variety of cell types such as embryonic fibroblasts, synovial cells, A549 cells, macrophages, endothelial cells, chondrocytes, smooth muscle cells, cardiac myocytes, and osteoblasts[18, 21-26]. *In vivo* studies also show that mPGES-1 is induced in animal models of inflammation such as the full thickness incisional model of wound healing[27], collagen-induced arthritis (CIA)[28], LPS-induced pyresis and adjuvant-induced arthritis[29-30]. mPGES-1 is dramatically upregulated in coordination with COX-2 in synovial tissue and in response to pro-inflammatory stimuli in synovial fibroblasts from RA patients[25]. Similarly, high levels of mPGES-1 are detected in human cartilage from OA patients and chondrocytes stimulated with pro-inflammatory cytokines[31-32].

Studies using mPGES-1 knockout (KO) mice have revealed the pathophysiological role of mPGES-1 in inflammatory diseases. One of the first reports using mPGES-1 KO mice subjected to subcutaneous administration of native chicken collagen type II (CIA model of arthritis in mice) showed that mPGES-1 KO mice exhibited a significant reduction in the incidence of arthritis, inflammation and histopathological changes compared to WT mice[33]. In a subsequent report, mPGES-1 KO CIA mice subjected to injection of bovine collagen type II showed that mPGES-1 null and heterozygous mice exhibited decreased incidence and severity of arthritis associated with attenuation in mechanical nociception and marked reduction of serum immunoglobulins (IgG) against type II collagen, including subclasses IgG1, IgG2a, IgG2b, IgG2c, and IgG3, compared to wild-type mice[28]. These results demonstrated that deficiency of mPGES-1 inhibits the development of CIA, at least in part, by blocking the development of a humoral immune response against type II collagen. Furthermore, in a collagen antibody-induced arthritis (CAIA) model, mPGES-1 KO mice displayed significant reduction in inflammation associated with reduced PGE_2 production[34].

The role of mPGES-1 in fever and pain has also been reported. Injection with IL-1β and LPS induces the expression of mPGES-1 along with COX-2 in blood brain barrier endothelial cells[35]. mPGES-1 KO mice subjected to LPS do not develop fever compared to WT mice[36]. This report also showed that mPGES-1 KO mice had normal levels of PGE_2 compared to high levels of PGE_2 in WT mice in response to LPS injection. Similarly, in a model of peripheral inflammation induced by turpentine, mPGES-1 KO did not exhibit febrile response while WT mice did [37].

COX-2 versus mPGES-1 in prostaglandin biosynthesis: Focus on Cardiovascular Effects

Inhibitors of COX-2 impart their anti-inflammatory effects and pain relief by inhibiting both PGE_2 and PGI_2. One of the hypotheses to explain the cardiovascular effects of COX-2 inhibition is the loss of anti-thrombotic PGI_2 derived from endothelial COX-2, which plays a key role in the regulation of thrombogenesis[17]. Animal studies using COX-2 KO mice, mice expressing modified COX-2 with no cyclooxygenase activity but intact peroxidase activity, and mice treated with celecoxib all show not only reduced PGE_2 production but also significant reduction in prostacyclin metabolite production with no change in thromboxane metabolite levels[38]. Furthermore, this study reported that mice with COX-2 genetic deletion, mutation or treatment with celecoxib end up developing thrombosis and hypertension. In contrast, mPGES-1 KO mice showed significant reduction in PGE_2 production but exhibited increased production of prostacyclin metabolites with no change in thromboxane metabolite levels. mPGES-1 KO mice fed with a normal or high salt diet failed to exhibit altered blood pressure or thrombogenesis. A follow up of this study showed that atherogenesis in mice lacking the receptor for low-density lipoprotein (LDL) exhibit upregulation of mPGES-1[39]. Further, genetic deletion of mPGES-1 retards atherogenesis with significant reduction in lesional macrophages and macrophage-derived foam cells associated with decreased PGE_2 production, and increased PGI_2 production with no change in thromboxane production.

Similar diversion in prostaglandin production profile as a result of genetic deletion of mPGES-1 has been reported in a variety of cell types including mouse embryonic fibroblasts (MEFs) and macrophages[22, 40]. Genetic deletion of COX-2 in mouse embryonic fibroblasts (MEFs) results in decreased production of both PGE_2 and 6-keto $PGF_{1\alpha}$ (the stable metabolic product of PGI_2) under basal conditions as well as after stimulation with IL-1β[22]. Genetic deletion of mPGES-1 in MEFs results in a dramatic decrease in PGE_2 production in mPGES-1 heterozygous (Het) and KO MEFs under basal conditions as well as after stimulation with IL-1β, suggesting that mPGES-1 is critically important for PGE_2 production[22-23]. Furthermore, mPGES-1 gene deletion results in diversion of prostanoid production from PGE_2 to 6-keto $PGF_{1\alpha}$ in mPGES-1 Het and KO MEFs compared to their WT MEFs, without altering the production profile of other prostaglandins. Similarly, in mouse peritoneal macrophages derived from mPGES-1 KO mice, PGE_2 was the predominant prostanoid produced in the LPS stimulated mPGES-1 WT mouse peritoneal macrophages. Incubation of mPGES-1 null mouse peritoneal macrophages with LPS resulted in a significant decrease in the levels of PGE_2 but significant increase in the levels of other prostanoids in the following order: TXB_2 > 6-keto-$PGF_{1\alpha}$ > $PGF_{2\alpha}$ > PGD_2 > PGE_2 [40]. These results indicate that in the absence of induced mPGES-1, elevation of COX-2 after inflammatory insult results in increase in the availability of PGH_2 as the common substrate for the generation of other prostanoids. Most importantly, the predominant product of PGH_2 shunting towards a preferred prostanoid varies from cell to cell.

Will Pharmacological Inhibition of mPGES-1 Counteract Inflammation and Escape Cardiovascular Side Effects Associated with COX-2 Inhibitors?

Is there any specific and potent inhibitor for mPGES-1? In 2005, MK866 (5 lipoxygenase activating protein indole inhibitor) was reported as an inhibitor of mPGES-1 with IC50 of 1.6 microM and was used as a lead structure to develop nanomolar inhibitors of mPGES-1[41]. These inhibitors exhibited strong protein binding properties and lacked potency. A recent study published by Merck Frosst Centre for Therapeutic Research (Canada) showed the discovery of Phenanthrene imidazoles 26 and 44 as potent, selective and orally active mPGES-1 inhibitors[42]. These inhibitors exhibit human whole blood IC50 of 0.20 and 0.14 microM, respectively, and were shown to exhibit a significant analgesic effect in a guinea pig hyperalgesia model. However, their effects in animal models of disease as well as toxicity data are yet to be reported.

Studies using mPGES-1 KO mice do suggest that mPGES-1 is an attractive target for counteracting inflammation. However, the major questions are: Will mPGES-1 pharmacological inhibition (1) mimic functional effects similar to its gene deletion, and (2) escape the cardiovascular side effects associated with COX-2 inhibitors? These questions will remain unanswered until a potent (specific) inhibitor of mPGES-1 is tested in animal studies and, subsequently, in a clinical setting. The fact that selective inhibition of COX-2 or genetic deletion of IP (the receptor for PGI_2) results in accelerated thrombogenesis and elevated blood pressure[38, 43-44] in contrast to genetic deletion of mPGES-1 which does not accelerate response to thrombogenesis[38], does not lead to hypertension and atherogenesis[39], attenuates abdominal aortic aneurysms[45] and exhibits reduced magnitude of infarction and tissue injury in transient focal ischemia *in vivo*[46], is very promising. However, the consequences of prostaglandin re-diversion as a result of mPGES-1 deletion (inhibition) on anti-inflammatory and cardiovascular effects raise few concerns. Firstly, the consequences of mPGES-1 genetic deletion and re-diversion of the prostanoid production profile from predominant PGE_2 to PGI_2[22, 38] needs careful scrutiny since mice lacking the IP receptor show that PGI_2 can itself mediate pain, inflammation and response to arthrogenic stimuli *in vivo*[38, 43-44]. Secondly, it is almost essential to establish an understanding of the re-diversion pattern for a specific PG product upon deletion (inhibition) of mPGES-1 in diseased cell types, as PG re-diversion towards a preferred PG product varies from cell type to cell type resulting in altered functions[22, 38, 40]. Thirdly, what are the consequences of specific inhibition of PGE_2 on cardiac and vascular functions, as PGE_2 mediates its effects through a complex network of four EP receptors (EP1-4) that mediate contrasting inflammatory and vascular function [47-51].

At present, it would be premature to predict that inhibition of mPGES-1 would yield desired results and this will certainly remain unanswered until a potent (specific) inhibitor of mPGES-1 is tested in experimental animals and hopefully in a clinical setting. The mystery of mPGES-1 inhibition and its therapeutic potential continues.

REFERENCES

[1] Vane, JR. Inhibition of prostaglandin synthesis as a mechanism of action for aspirin-like drugs. *Nat New Biol.* Jun 23, 1971, 231(25), 232-235.

[2] Kujubu, DA; Fletcher, BS; Varnum, BC; Lim, RW; Herschman, HR. TIS10, a phorbol ester tumor promoter-inducible mRNA from Swiss 3T3 cells, encodes a novel prostaglandin synthase/cyclooxygenase homologue. *J Biol Chem.,* Jul 15 1991, 266(20), 12866-12872.

[3] Kapoor, M; Kojima, F; LJ, C. Arachidonic acid-derived eicosanoids in rheumatoid arthritis: implications and future targets. *Future Rheumatology.* 2006, 1(3), 323-330.

[4] Xie, WL; Chipman, JG; Robertson, DL; Erikson, RL; Simmons, DL. Expression of a mitogen-responsive gene encoding prostaglandin synthase is regulated by mRNA splicing. *Proc Natl Acad Sci* U S A. Apr 1, 1991, 88(7), 2692-2696.

[5] Crofford, LJ. COX-1 COX-2 tissue expression: implications and predictions. *J Rheumatol Suppl.,* Jul 1997, 49, 15-19.

[6] Martel-Pelletier, J; Pelletier, JP; Fahmi, H. Cyclooxygenase-2 and prostaglandins in articular tissues. *Semin Arthritis Rheum.,* Dec 2003, 33(3), 155-167.

[7] Angel J, Berenbaum F, Le Denmat C, Nevalainen T, Masliah J, Fournier C. Interleukin-1-induced prostaglandin E2 biosynthesis in human synovial cells involves the activation of cytosolic phospholipase A2 and cyclooxygenase-2. *Eur J Biochem.* Nov 15 1994, 226(1), 125-131.

[8] Crofford LJ, Wilder RL, Ristimaki AP, et al. Cyclooxygenase-1 and -2 expression in rheumatoid synovial tissues. Effects of interleukin-1 beta, phorbol ester, and corticosteroids. *J Clin Invest.* Mar 1994, 93(3), 1095-1101.

[9] Hulkower KI; Wertheimer SJ; Levin W; et al. Interleukin-1 beta induces cytosolic phospholipase A2 and prostaglandin H synthase in rheumatoid synovial fibroblasts. Evidence for their roles in the production of prostaglandin E2. *Arthritis Rheum.* May 1994, 37(5), 653-661.

[10] Kawai S; Nishida S; Kato M; et al. Comparison of cyclooxygenase-1 and -2 inhibitory activities of various nonsteroidal anti-inflammatory drugs using human platelets and

synovial cells. *Eur J Pharmacol.* Apr 17 1998, 347(1), 87-94.

[11] Crofford LJ. COX-2 in synovial tissues. *Osteoarthritis Cartilage.* Jul 1999, 7(4), 406-408.

[12] Pelletier JP; Martel-Pelletier J. Therapeutic targets in osteoarthritis: from today to tomorrow with new imaging technology. *Ann Rheum Dis.* Nov 2003, 62 Suppl 2, ii79-82.

[13] Meade EA; Smith WL; DeWitt DL. Differential inhibition of prostaglandin endoperoxide synthase (cyclooxygenase) isozymes by aspirin and other non-steroidal anti-inflammatory drugs. *J Biol Chem.* Mar 25 1993, 268(9), 6610-6614.

[14] Bombardier C; Laine L; Reicin A; et al. Comparison of upper gastrointestinal toxicity of rofecoxib and naproxen in patients with rheumatoid arthritis. VIGOR Study Group. *N Engl J Med.* Nov 23 2000, 343(21), 1520-1528, 1522 p following 1528.

[15] Bresalier RS; Sandler RS; Quan H; et al. Cardiovascular events associated with rofecoxib in a colorectal adenoma chemoprevention trial. *N Engl J Med.* Mar 17 2005, 352(11), 1092-1102.

[16] Solomon SD; McMurray JJ; Pfeffer MA; et al. Cardiovascular risk associated with celecoxib in a clinical trial for colorectal adenoma prevention. *N Engl J Med.* Mar 17 2005, 352(11), 1071-1080.

[17] McAdam BF; Catella-Lawson F; Mardini IA; Kapoor S; Lawson JA; FitzGerald GA. Systemic biosynthesis of prostacyclin by cyclooxygenase (COX)-2: the human pharmacology of a selective inhibitor of COX-2. *Proc Natl Acad Sci U S A.* Jan 5 1999, 96(1), 272-277.

[18] Jakobsson PJ; Thoren S; Morgenstern R; Samuelsson B. Identification of human prostaglandin E synthase: a microsomal, glutathione-dependent, inducible enzyme, constituting a potential novel drug target. *Proc Natl Acad Sci U S A.* Jun 22 1999, 96(13), 7220-7225.

[19] Tanikawa N; Ohmiya Y; Ohkubo H; et al. Identification and characterization of a novel type of membrane-associated prostaglandin E synthase. *Biochem Biophys Res Commun.* Mar 8 2002, 291(4), 884-889.

[20] Tanioka T; Nakatani Y; Semmyo N; Murakami M; Kudo I. Molecular identification of cytosolic prostaglandin E2 synthase that is functionally coupled with cyclooxygenase-1 in immediate prostaglandin E2 biosynthesis. *J Biol Chem.* Oct 20 2000, 275(42), 32775-32782.

[21] Murakami M; Naraba H; Tanioka T; et al. Regulation of prostaglandin E2 biosynthesis by inducible membrane-associated prostaglandin E2 synthase that acts in concert with cyclooxygenase-2. *J Biol Chem.* Oct 20 2000, 275(42), 32783-32792.

[22] Kapoor M; Kojima F, Qian M; Yang L; Crofford LJ. Shunting of prostanoid biosynthesis in microsomal prostaglandin E synthase-1 null embryo fibroblasts: regulatory effects on inducible nitric oxide synthase expression and nitrite synthesis. *Faseb J.* 2006, 20(13), 2387-2389.

[23] Kapoor M; Kojima F; Qian M; Yang L; Crofford LJ. Microsomal prostaglandin E synthase-1 deficiency is associated with elevated peroxisome proliferator activated receptor gamma: Regulation by prostaglandin E2 via the PI3 kinase and AKT pathway. *J Biol Chem.* Dec 22 2006.

[24] Kojima F; Naraba H; Sasaki Y; Beppu M; Aoki H; Kawai S. Prostaglandin E2 is an enhancer of interleukin-1beta-induced expression of membrane-associated

prostaglandin E synthase in rheumatoid synovial fibroblasts. *Arthritis Rheum.* Oct 2003, 48(10), 2819-2828.

[25] Kojima F; Naraba H; Sasaki Y; Okamoto R; Koshino T; Kawai S. Coexpression of microsomal prostaglandin E synthase with cyclooxygenase-2 in human rheumatoid synovial cells. *J Rheumatol.* Sep 2002, 29(9), 1836-1842.

[26] Stichtenoth DO; Thoren S; Bian H; Peters-Golden M; Jakobsson PJ; Crofford LJ. Microsomal prostaglandin E synthase is regulated by proinflammatory cytokines and glucocorticoids in primary rheumatoid synovial cells. *J Immunol.* Jul 1 2001, 167(1), 469-474.

[27] Kapoor M; Kojima F; Yang L; Crofford LJ. Sequential induction of pro- and anti-inflammatory prostaglandins and peroxisome proliferators-activated receptor-gamma during normal wound healing: A time course study. *Prostaglandins Leukot Essent Fatty Acids.* Jan 17 2007.

[28] Kojima F; Kapoor M; Yang L; et al. Defective generation of a humoral immune response is associated with a reduced incidence and severity of collagen-induced arthritis in microsomal prostaglandin E synthase-1 null mice. *J Immunol.* Jun 15 2008, 180(12), 8361-8368.

[29] Claveau D; Sirinyan M; Guay J; et al. Microsomal prostaglandin E synthase-1 is a major terminal synthase that is selectively up-regulated during cyclooxygenase-2-dependent prostaglandin E2 production in the rat adjuvant-induced arthritis model. *J Immunol.* May 1 2003, 170(9), 4738-4744.

[30] Mancini JA; Blood K; Guay J; et al. Cloning, expression, and up-regulation of inducible rat prostaglandin e synthase during lipopolysaccharide-induced pyresis and adjuvant-induced arthritis. *J Biol Chem.* Feb 9 2001, 276(6), 4469-4475.

[31] Li X; Afif H; Cheng S; et al. Expression and regulation of microsomal prostaglandin E synthase-1 in human osteoarthritic cartilage and chondrocytes. *J Rheumatol.* May 2005, 32(5), 887-895.

[32] Kojima F; Naraba H; Miyamoto S; Beppu M; Aoki H; Kawai S. Membrane-associated prostaglandin E synthase-1 is upregulated by proinflammatory cytokines in chondrocytes from patients with osteoarthritis. *Arthritis Res Ther.* 2004, 6(4), R355-365.

[33] Trebino CE; Stock JL; Gibbons CP; et al. Impaired inflammatory and pain responses in mice lacking an inducible prostaglandin E synthase. *Proc Natl Acad Sci U S A.* Jul 22 2003, 100(15), 9044-9049.

[34] Kamei D; Yamakawa K; Takegoshi Y; et al. Reduced pain hypersensitivity and inflammation in mice lacking microsomal prostaglandin e synthase-1. *J Biol Chem.* Aug 6 2004, 279(32), 33684-33695.

[35] Ek M; Engblom D; Saha S; Blomqvist A; Jakobsson PJ; Ericsson-Dahlstrand A. Inflammatory response: pathway across the blood-brain barrier. *Nature.* Mar 22 2001, 410(6827), 430-431.

[36] Engblom D; Saha S; Engstrom L; et al. Microsomal prostaglandin E synthase-1 is the central switch during immune-induced pyresis. *Nat Neurosci.* Nov 2003, 6(11), 1137-1138.

[37] Saha S; Engstrom L; Mackerlova L; Jakobsson PJ; Blomqvist A. Impaired febrile responses to immune challenge in mice deficient in microsomal prostaglandin E synthase-1. *Am J Physiol Regul Integr Comp Physiol.* May 2005, 288(5), R1100-1107.

[38] Cheng Y; Wang M; Yu Y; Lawson J; Funk CD; Fitzgerald GA. Cyclooxygenases, microsomal prostaglandin E synthase-1, and cardiovascular function. *J Clin Invest.* May 2006, 116(5), 1391-1399.

[39] Wang M; Zukas AM; Hui Y; Ricciotti E; Pure E; FitzGerald GA. Deletion of microsomal prostaglandin E synthase-1 augments prostacyclin and retards atherogenesis. *Proc Natl Acad Sci U S A.* Sep 26 2006, 103(39), 14507-14512.

[40] Trebino CE; Eskra JD; Wachtmann TS; Perez JR; Carty TJ; Audoly LP. Redirection of eicosanoid metabolism in mPGES-1-deficient macrophages. *J Biol Chem.* Apr 29 2005, 280(17), 16579-16585.

[41] Riendeau D; Aspiotis R; Ethier D; et al. Inhibitors of the inducible microsomal prostaglandin E2 synthase (mPGES-1) derived from MK-886. *Bioorg Med Chem Lett.* Jul 15 2005, 15(14), 3352-3355.

[42] Giroux A; Boulet L; Brideau C; et al. Discovery of disubstituted phenanthrene imidazoles as potent, selective and orally active mPGES-1 inhibitors. *Bioorg Med Chem Lett.* Oct 15 2009, 19(20), 5837-5841.

[43] Murata T; Ushikubi F; Matsuoka T; et al. Altered pain perception and inflammatory response in mice lacking prostacyclin receptor. *Nature.* Aug 14 1997, 388(6643), 678-682.

[44] Honda T; Segi-Nishida E; Miyachi Y; Narumiya S. Prostacyclin-IP signaling and prostaglandin E2-EP2/EP4 signaling both mediate joint inflammation in mouse collagen-induced arthritis. *J Exp Med.* Feb 20 2006, 203(2), 325-335.

[45] Wang M; Lee E; Song W; et al. Microsomal prostaglandin E synthase-1 deletion suppresses oxidative stress and angiotensin II-induced abdominal aortic aneurysm formation. *Circulation.* Mar 11 2008, 117(10), 1302-1309.

[46] Ikeda-Matsuo Y; Ota A; Fukada T; Uematsu S; Akira S; Sasaki Y. Microsomal prostaglandin E synthase-1 is a critical factor of stroke-reperfusion injury. *Proc Natl Acad Sci U S A.* Aug 1 2006, 103(31), 11790-11795.

[47] Takayama K; Garcia-Cardena G; Sukhova GK; Comander J; Gimbrone MA, Jr.; Libby P. Prostaglandin E2 suppresses chemokine production in human macrophages through the EP4 receptor. *J Biol Chem.* Nov 15 2002, 277(46), 44147-44154.

[48] Ushikubi F; Segi E; Sugimoto Y; et al. Impaired febrile response in mice lacking the prostaglandin E receptor subtype EP3. *Nature.* Sep 17 1998, 395(6699), 281-284.

[49] Kennedy CR; Zhang Y; Brandon S; et al. Salt-sensitive hypertension and reduced fertility in mice lacking the prostaglandin EP2 receptor. *Nat Med.* Feb 1999, 5(2), 217-220.

[50] Takayama K; Sukhova GK; Chin MT; Libby P. A novel prostaglandin E receptor 4-associated protein participates in antiinflammatory signaling. *Circ Res.* Mar 3 2006, 98(4), 499-504.

[51] Norel X; de Montpreville V; Brink C. Vasoconstriction induced by activation of EP1 and EP3 receptors in human lung, effects of ONO-AE-248, ONO-DI-004, ONO-8711 or ONO-8713. *Prostaglandins Other Lipid Mediat.* Oct 2004, 74(1-4), 101-112.

In: Prostaglandins: Biochemistry, Functions… ISBN: 978-1-61668-272-9
Editors: Gillian M. Goodwin, pp.113-126

Chapter 8

INVESTIGATION OF KINETICS OF BEHAVIOR OF STRUCTURES OF PROSTAGLANDINS OF GROUPS A_1, A_2, B_1, B_2 AND E_1, E_2 IN ETHANOL. PROSTAGLANDIN INTERCHANGE

S. V. Isay, T. V. Kafanova and N. U. Kim
Pacific institute of bioorganic chemistry Far East Division of Russian Academy Sciences, Vladivostok 690022, Prospekt 100-letiya of Vladivostok, 159

ABSTRACT

The kinetic behavior of structures of prostaglandins A_1, A_2, B_1, B_2 and E_1, E_2 in ethanol is explored at 20-25°C by a method of a UV-spectroscopy. The absorbance readings were performed monthly within one year. It is shown, that prostaglandins are interchanged and the character of interchanges is individual. So, partial conversion of E_1 to A_1 needs one year. The conversion of E_2 to A_2 begins in three months. The intensive conversion of $E_2 \rightarrow B_2$ begins in one month, and in 12 months the equilibrium between E_2 and B_2 is established practically. At the same time reversible conversion of $B_2 \rightarrow E_2$ begins in 1 month, some more in 11 months, along with B_2, we have intensive peaks A_2 and E_2.

INTRODUCTION

Earlier we have shown, that the basic groups of prostaglandins (PGs) in composition of an extract of a soft coral *Plexaura homomalla* preserve their structures and biological activity within many years [1]. This fact contradicted the settled opinion on an extraordinary PG instability. Probably answer to a question of PG stability is consisted in properties of PGs. It is necessary to note, that in the literature of works on investigation of PG properties are unities [2-9], at the same time works in this direction are important both for practice, and theory of prostaglandinology .

The purpose of our investigation - to study kinetics of behavior of structures of the PG standards in ethanol at room temperature.

METHODS

Materials

We had PG samples of groups A and B of the two series of firm Serva (Germany) and E_1, E_2 (factory of biological preparations, Tallin, Estonia). PG samples A_1 and B_1 in ampoules were crystalline; PG samples A_2 and B_2 were as oil; E_1 and E_2 were crystalline. After opening ampoules with PG samples, stored at -20°C, ethanol solutions at concentrations: A_1, A_2, B_1, B_2 – 1 mg in 100 μl were prepared. For spectral scan aliquots on 5 μl were selected. Initial concentrations of E_1 and E_2 were 5 mg/ml. For spectral scan aliquots on 20 μl were selected. Samples further were stored on a laboratory table at ambient temperature (20-25°C).

Equipment

Spectral scans for PGs have been performed on the device spectrophotometer Cecil CE 7200 (Aquarius, England) monthly within one year. Quartz cells were used with 1-cm path-length and 1-ml volume. An interval for scaning is 200-400 nm.

Table 1. Molar extinction coefficients (ε) of starting ethanol solutions of the PG standards at ambient temperature (20- 25°C) at various wavelengths

Wavelength (nm):	201	202	203	204	205	207	208	217	271	278	280
Compounds						coefficients of a molar extinction (ε)					
A_1		13.988	16.966		13.041			13.414		10.778	
A_2	5..580					6..520			4..250		
E_1							2.481				
E_2					4.400						
B_1				3.162						10.953	
B_2	4..997			4..976							9.706

RESULTS

After opening ampouls with PG samples and the preparations from them of ethanol solutions with concrete concentration UV spectra for each of the solutions prepared were performed at wavelength interval 200-400 nm.

After obtaining of spectra for everyone PG the coefficients of a molar extinction (ε) are calculated. The results are submitted in the Table 1. Further spectral scans are yielded monthly within one year. The exception was made some by cases for technical reasons. The spectra for some PGs in dynamic are submitted in the figures 1-4. On spectra for everyone PG the values ε are calculated. They are submitted in the tables 2-4.

DISCUSSION

It is known, that PGs of groups A, B, E, have characteristic absorbances in UV field, therefore main method of the analysis in examination of kinetics of change of PG structures in ethanol is chosen UV spectroscopy. UV spectra of solutions of the PG standards, prepared immediately after opening ampoules, have shown, that each of PGs, except for E_1 and E_2, had a few absorbance peaks (table 1), that corresponded literature data [9]. Earlier in one of our work [10], performing the mass spectrometric (MS) analysis of the PG standards, we have found out, that, alongside with the main peaks, characteristic for concrete PG, there are also peaks, the value m/z which is much less, but which can be related to other PG types. UV spectroscopy also gives for individual PGs a few peaks at different wavelengths [9]. It is possible to suppose, that the PG standards or are initially polluted with each other at their obtaining in industrial conditions, or PGs are capable to an isomerization. PGs are easily isomerizated under action of acids and alkalis, fact known [2, 8, 11, 12]. The conditions of carrying out of reactions in these investigations are various. The authors use wide intervals pH and media, in which observe kinetics: aqueous, various solvents, buffered solutions, aqueous-alcoholic solutions, emulsions; a various temperature regime [2, 5-9, 12, 13]. We have met only three works about behavior of PGs in absence of acids and alkalis [3, 4, 13].

We explored kinetics of behavior 6 PGs of two series (first and second) in an ethanol solution at room temperature (20-25°C) within one year. The choice of an ethanol is not accidental. In the literature there are references that in case of use of PGE solutions for clinical practice, they should be stored as concentrated alcoholic solutions in ampoules [2-4]. At the same time it should be noted that ethanol has a greater acid properties than methanol and ethylene glycol. Data obtained, submitted in figures and tables, show, that PGs are interchanged and the character of interchanges is individual. So, conversion of a part of E_1 to A_1 needs 1 year, the conversion of $E_2 \rightarrow A_2$ begins in 3 months. Thereat in a case E_1 do not observe formation B_1, whereas in a case E_2, alongside with A_2, is formed B_2. Moreover, the partial conversion of $E_2 \rightarrow B_2$ occurs for 1 month. At the same time reverse conversion of $B_2 \rightarrow E_2$ demands 5 months. The conversion of B_2 to A_2 is stretched with time: from 3 about 12 months (table 4). The ε values, calculated for them, are agreed with literature data [9].

The kinetic courses of PG A_1 and A_2 strongly differ. So, the starting solution A_1 has sharp peak at 218 nm and simultaneously peak relative to B_1, at 278 nm almost in equilibrium (figure 1). At the same time for A_2 we have in a starting solution three peaks, but their position concerning standards for PG wavelengths is shifted. In the literature we met mention similar shifts. The authors explain their by different reasons. So, Monkhouse et al. [2] speaks that PGA at 218 nm frequently is screened, that we observe at itself for A_2. Amin [14] determined A_2 at 224 nm. Some authors explain shift in wavelengths by formation of intermediate compounds (keto- or hydroperoxides) [2, 9, 12]. In our investigation we also observed shifts of peaks, however, our attempts to identify with the help MS PG-keto (for example, 15-keto-PGE$_2$) or peroxide compounds have not given results (unpublished data).

Probably this question could be solved, using IR-spectroscopy, but the quantities of PGs with which we worked, did not allow it to make.

Prostaglandin A_1

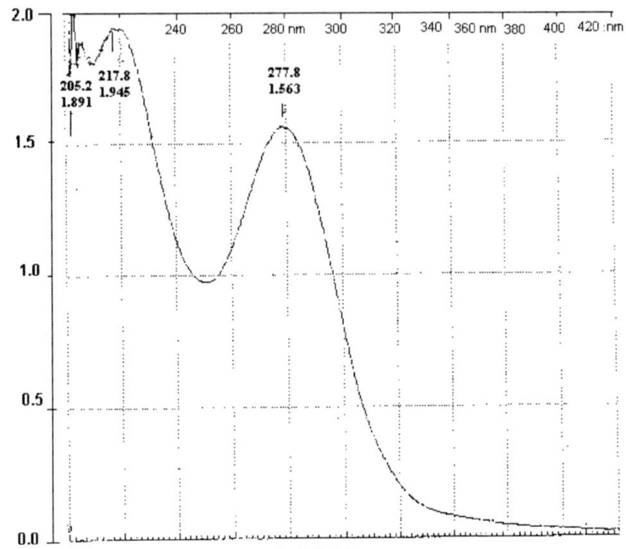

initial*

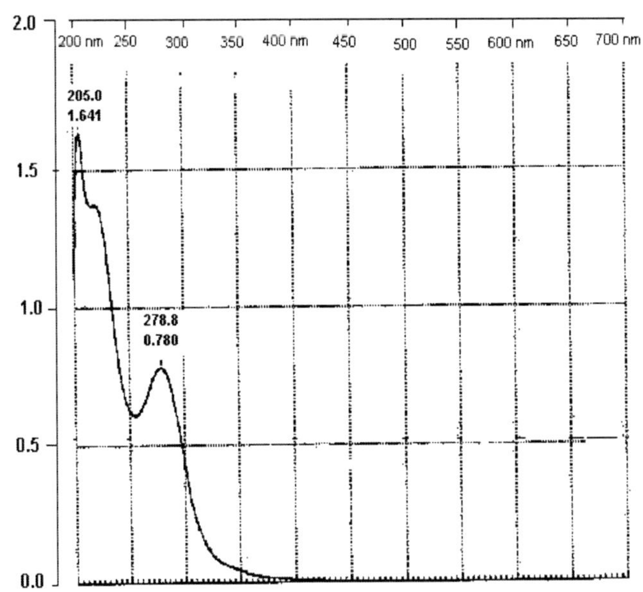

in 7 months

Figure 1 (continues).

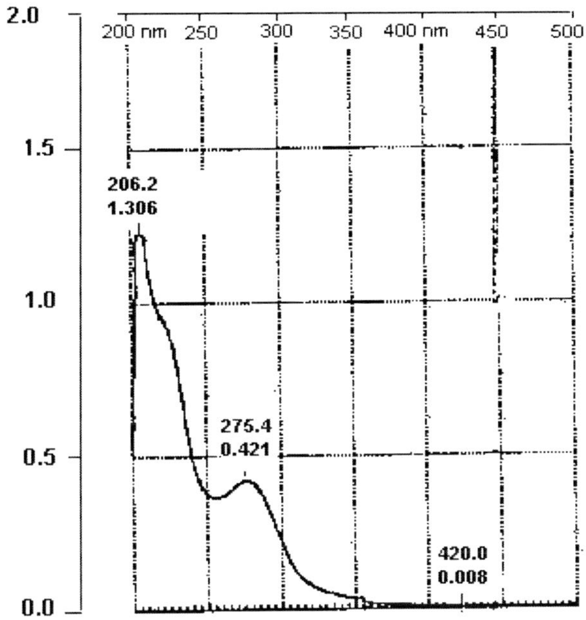

in 12 months

*- "initial"- it is necessary to understand as time of screen of a spectrum at once after opening of an ampoule with a PG sample

Figure 1. Kinetic of behavior of prostaglandin A_1 in time (months)

Prostaglandin B_1

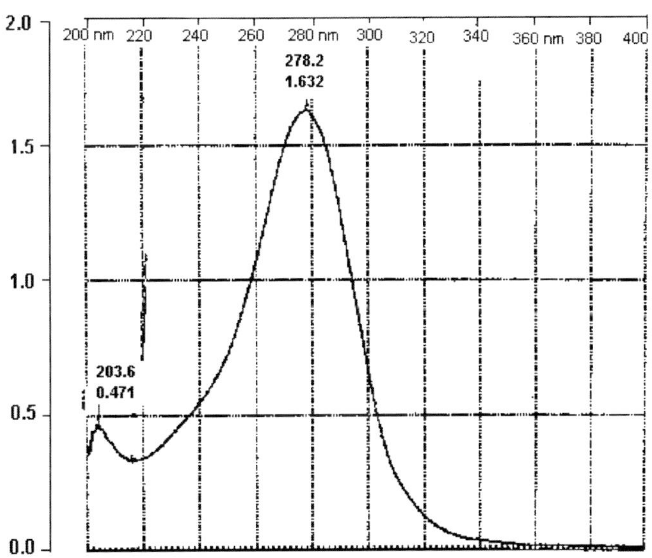

initial*

Figure 2 (continues).

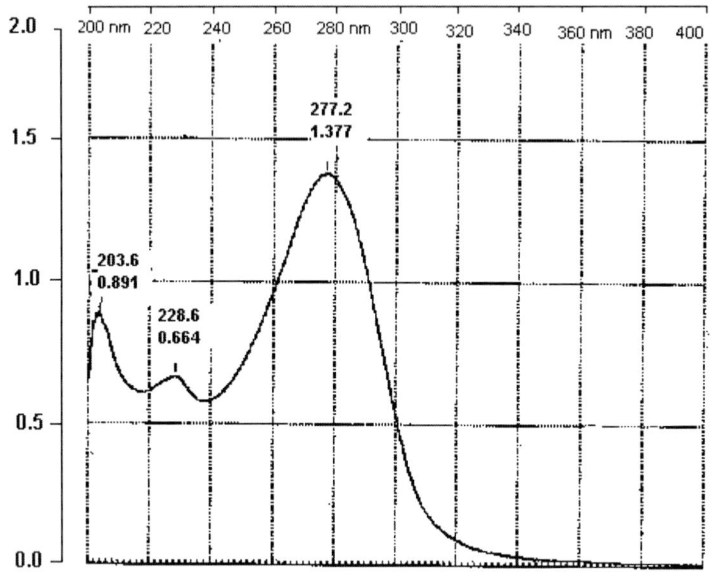

in 4 months

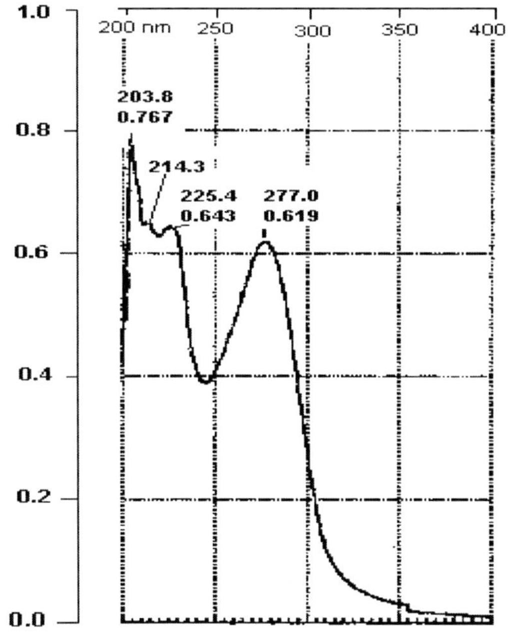

in12 months

*- "initial"- it is necessary to understand as time of screen of a spectrum at once after opening of an ampoule with a PG sample

Figure 2. Kinetic of behavior of prostaglandin B_1 in time (months)

Table 2. Molar extinction coefficients (ε) of ethanol solutions of PG A_1 and A_2 at temperature 20-25°C at various wavelengths in dynamics

A_1																
Wavelength (nm):	200	201	202	203	204	205	206	207	217.8	218	270	271	274	275	278	420
Time							coefficients of a molar extinction (ε)									
interval, month																
Starting solution			13.988	16.966		13.041			13.414						10.778	
1		10.152					12.446			12.237					8.586	
3						11.112				10.916					7.050	
4					11.400										6.149	
5		13.517	13.846												5.774	
8						11.317									5.379	
10	15.400			21.944											6.994	
12							18.014							5.806		
A_2																
Starting solution		5.580						6.520				4.250				
1							6.400						3.640			
3						3.300				5.690						
4			2.070		5.030											
5						7.220										
8				3.350												
10				5.427												
12	6.600					9.747										0.080

Table 3. Molar extinction coefficients (ε) of ethanol solutions of PG E_1 and E_2 at temperature 20-25°C at various wavelengths in dynamics

E_1																	
Wavelength (nm):	200	201	202	203	204	205	206	208	214	225	226	227	277	278	279	355	418
Time																	
interval, months							coefficients of a molar extinction (ε)										
Starting solution								2.481									
1								2.738									
3					3.257		3.380										
4								2.989								0.029	
5						3.856											
8					4.387												
10							5.060										
12		3.243	3.396	3.500	4.278	4.919			4.811								0.034
E_2																	
Starting solution						4.400											
1						5.514							0.962				
3							6.655				6.624			3.680			
4		6.510		5.633								6.520		6.170			
5			9.190			8.926				7.600				8.285			
8			8.270	8.888	8.250								9.440				
10			9.160		9.400									9.780			
12	6.387				10.479	9.908									10.464		1.330

Table 4. Molar extinction coefficients (ε) of ethanol solutions of PG B_1 and B_2 at temperature 20-25°C at various wavelengths in dynamics.

								B_1									
Wavelength: (nm):	200	201	202	203	204	205	206	214	225	227	229	274	276	277	278	279	280
Time interval, months						coefficients of a molar extinction (ε)											
Starting solution					3.162										10.953		
1						4 926					3.664				10.289		
3				5.173							3.520			8.160			
4					5.980						4.456			9.242			
5					8.067				5.563					7.830			
8		11.300		12.309						7.886				9.752			
10					14.886				9.570					9.342			
12	6.953				10.295	10.403									8.309		
								B_2									
Starting solution		4.997			4.976												9.706
1							6.600									9.240	
3		8.572		9.856							7.155		9.100				
4			7.100								7.094	8.288					
5	9.652				13.886							8.559					
8			15.096	11.097	15.573							8.653					
10					18.453							8.434					
12	12.345				13.822			12.153				7.561					

Prostaglandin B_2

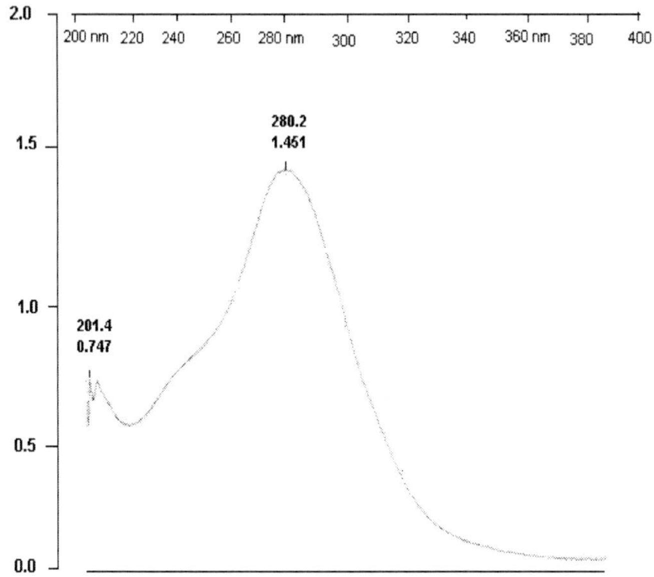

initial*

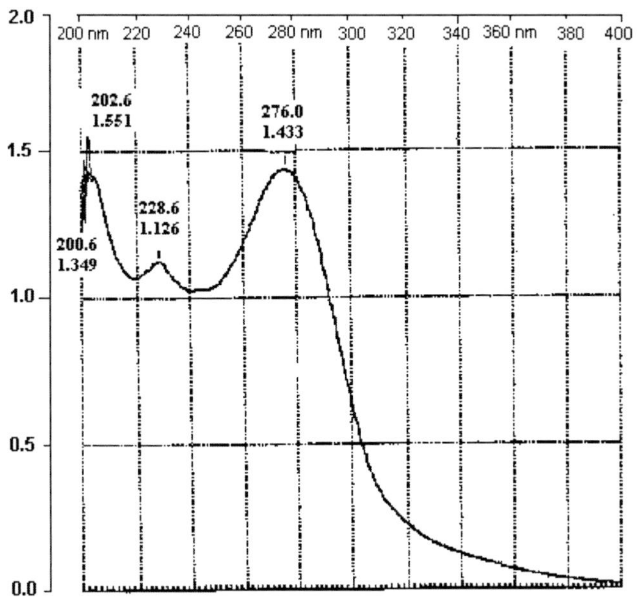

in 2,5 months

Figure 3 (continues).

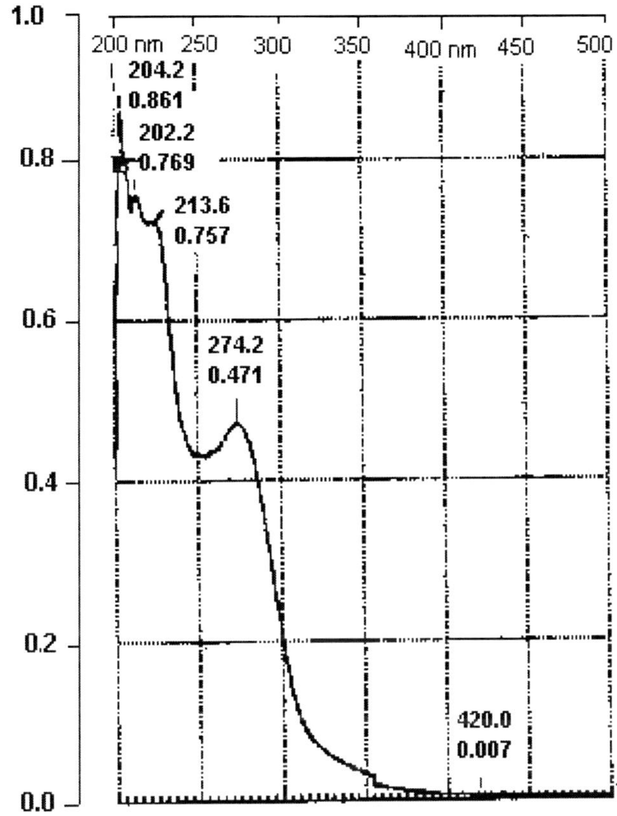

in 12 months

* - "initial"- it is necessary to understand as time of screen of a spectrum at once after opening of an ampoule with a PG sample

Figure 3. Kinetic of behavior of prostaglandin B_2 in time (months)

More often in the literature there are examinations on stability PGs of group E. So, Srivastava and Clausen have shown, that PG E_1, E_2, E_3 are stable in ethanol, chloroform and ethylacetate within 5 months [4]. Yalkovsky and Roseman [5] have shown, that the E-type prostaglandins are stable at 25°C in triacetin (acetate derivative of glycerol, dissolved in water and organic solvents) within 12 months. Our data on E_1 and E_2 in ethanol testify that in 12 months E_1 partially converses to A_1, and E_2 for the same period from an individual state passes practically in equilibrium with A_2 and B_2, that is we have equilibrium between $A_2 \leftrightarrow E_2 \leftrightarrow B_2$ (Table 3).

Alongside with examination of kinetics of behavior of individual PG structures, we have carried out similar investigation for their mixtures, separately first a series PGs (A_1, B_1, E_1) and second a series (A_2, B_2, E_2), varying their concentrations. PG UV spectra of first series were more sharp. In both series of PG original mixtures we observed the peaks of PGs E and B, peak, characteristic for PGA (217 nm) was screened. But after 1 year this peak is appeared sharply at 217 nm only in first series (A_1). UV spectra allow to see changes occurring with PGs in composition of mixtures.

Prostaglandin E_1

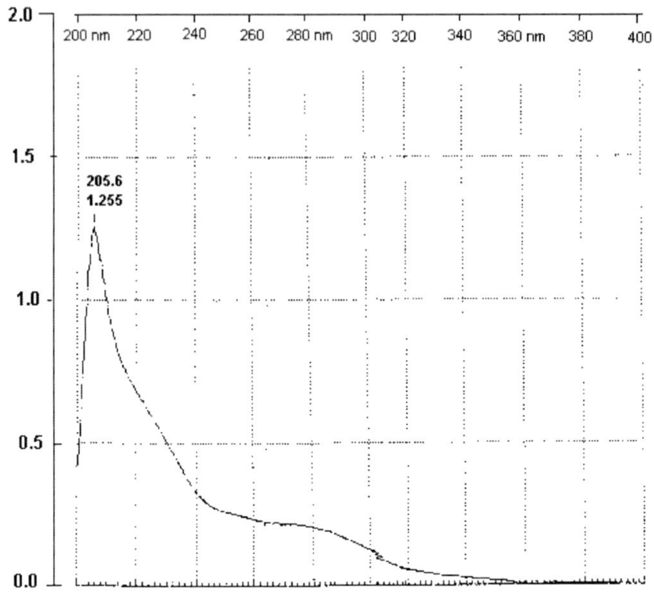

initial*

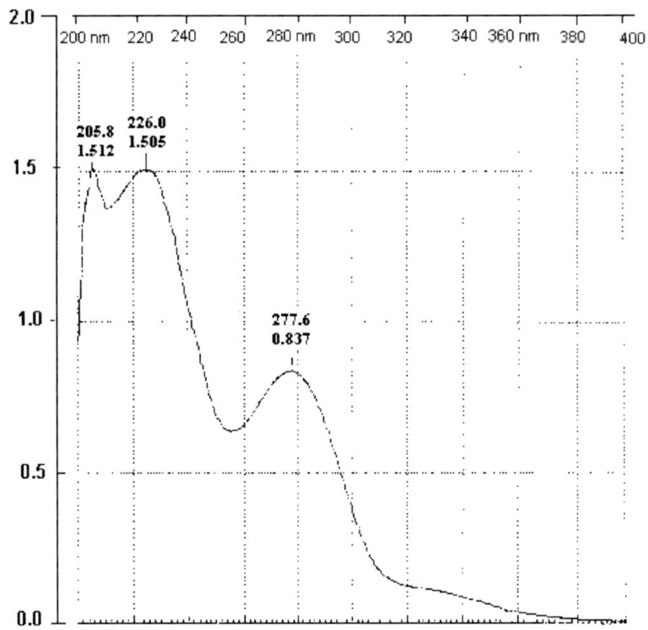

in 2,5 months

Figure 4 (continues).

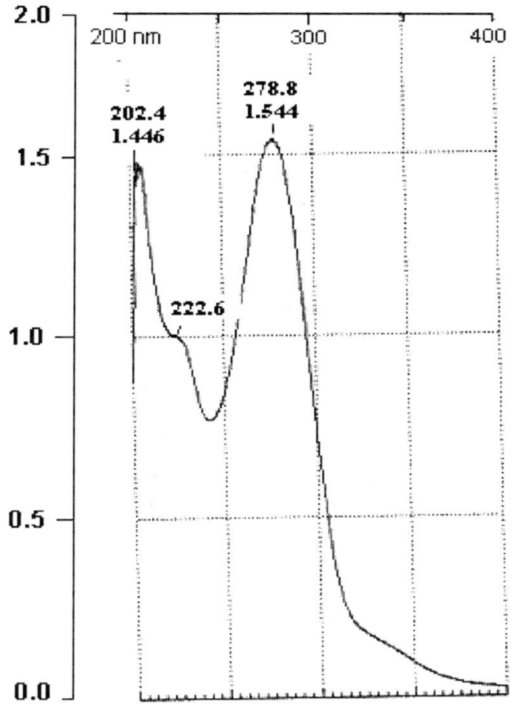

in 10 months

* - "initial"- it is necessary to understand as time of screen of a spectrum at once after opening of an ampoule with a PG sample

Figure 4. Kinetic of behavior of prostaglandin E_1 in time (months)

Results of examination on interchange of PG groups A, B, E could be expressed by the following cycle:

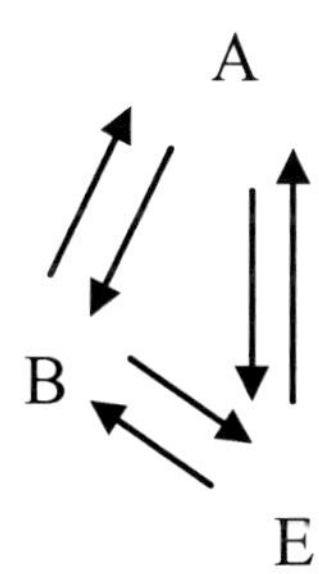

In our opinion, though we have no data on kinetic of behavior of PGF, in this "turnover" takes part and PGF. Indirect confirmation to this is ours early data [1] on examination of PG stability in *P. homomalla*, according to which we found out PGF in composition of an extract after 10 years of its storage.

Thus, being reverted to a start of article, it is possible to contend, that the long-time PG stability in composition of the *P. homomalla* extract first of all depends on PG ability to isomerizate each other.

References

[1] Isay, SV; Kafanova, TV; Zviagintseva, TYa. Prostaglandins of marine invertebrates-II. A qualitative assessment of prostaglandin stability in the soft coral *Plexaura homomalla* extract during long-time storage. *Comp. Biochem. Physiol.,* 1994, 109C, 205-214.

[2] Monkhouse, DC; Van Campen, L; Aguiar, AJ. Kinetics of dehydration and isomerization of prostaglandins E_1 and E_2. *J. Pharm. Sci.,* 1973, 62, 576-580.

[3] Roseman, TJ; Sims, B; Stehle, RG. Stability of prostaglandins. *Am. J. Hosp. Pharm.,* 1973, 30, 236-239.

[4] Srivastava, KC; Clausen, J. Stability of prostaglandin E compounds in solution. *Lipids,* 1973, 8, 592-594.

[5] Yalkovsky, SH; Roseman, TJ. Stability of E-type prostaglandins in triacetin. *J. Pharm. Sci.,* 1979, 68, 4214-4215.

[6] Stehle, RG. Prostaglandins and arachidonate metabolites. In: Lands WEM, Smith WL editors. *Methods in Enzymology*. New York, London: Academic Press, 1982, 436-458.

[7] Younger, EW; Szabo, RM. The stability of prostaglandin E_1 in dilute physiological solutions at 37 degrees C. *Prostaglandins,* 1986, 31, 923-927.

[8] Teagarden, DL; Anderson, BD; Petre, WJ. Dehydration kinetics of prostaglandin E_1 in a lipid emulsion. *Pharm. Res.,* 1989, 6, 210-215.

[9] Terragno, A; Rydzik, R; Terragno, NA. High performance liquid chromatography and UV detection for the separation and quantitation of prostaglandins. *Prostaglandins,* 1981, 21, 101-112.

[10] Isay, SV; Kafanova, TV; Matveeva, ER; Mirgorodskaya, OA. Mass spectrometry with extraction of dissolved ions at atmospheric pressure for prostaglandin identification in biological materials. *Org. Mass Spectrom.,* 1993, 28, 279-281.

[11] Andersen, NH. Dehydration of prostaglandins: study by spectroscopic method. *J. Lipid Res.*, 1969, 10, 320-325.

[12] Perera, SK; Fedor, LR. Acid- and base-catalyzed dehydration of prostaglandin E_2 to prostaglandin A_2 and general-base-catalyzed isomerization of prostaglandin A_2 to prostaglandin B_2. *J. Am. Chem. Soc.,* 1979, 101, 7390-7393.

[13] Muriel, P; Razzouq, N; Pharm, M; Tixer, G; Astier, A. Stability of prostaglandin E_1 (PGE_1) in aqueous solutions. *Eur. J. Hospital Pharm. Sci.,* 2006, 11, 31-36.

[14] Amin, M. Direct quantitative thin-layer chromatographic determination of prostaglandins A_2, D_2, E_2 and $F_{2\alpha}$. *J. Chromatogr.,* 1975, 108, 313-321.

In: Prostaglandins: Biochemistry, Functions...
Editors: Gillian M. Goodwin, pp.127-133
ISBN: 978-1-61668-272-9

Chapter 9

ROLES OF PROSTAGLANDIN E2 AND ITS RECEPTORS IN INFLAMMATION-RELATED CARDIOVASCULAR DISEASES

Jun-ichi Suzuki[1*], Masahito Ogawa[1], Keiichi Hishikari[2], Kiyoshi Takayama[3] and Mitsuaki Isobe[2]
[1]Department of Advanced Clinical Science and Therapeutics, University of Tokyo, Japan
[2]Department of Cardiovascular Medicine, Tokyo Medical and Dental University, Japan
[3]NB Health Laboratory, Japan

ABSTRACT

Prostaglandin E2 (PGE_2) is produced during inflammatory responses and controls a variety of immunological reactions through four different receptor subtypes (EP1-4) with various signaling cascades. However, the precise roles of each receptor in cardiovascular diseases have not yet been elucidated. It has been shown that PGE_2 suppresses myocardial ischemia and heart transplant rejection via its receptor EP4, suggesting a particular role of this subtype for myocardial inflammation. It has also been reported that enhanced expression of some subtypes of PGE_2 receptors has been shown in clinical and experimental cardiovascular diseases. However, their pathophysiological roles in the development of disorders remain unelucidated. Recently, we revealed the significance of EP4 for suppression of myocardial inflammation using a novel EP4 selective agonist (EP4RAG) in murine cardiac transplantation and myocardial ischemia models. The investigation indicated that EP4RAG attenuated inflammation by suppressing chemokines and the infiltration of macrophages. In this short review article, we are focusing on the roles of prostaglandin E2 receptors in the development of cardiovascular diseases.

---

* Corresponding author: Department of Advanced Clinical Science and Therapeutics, University of Tokyo, 7-3-1 Hongo, Bunkyo, Tokyo 113-8655, Japan. Phone, 81-3-5800-9116; Fax, 81-3-5800-9182; E-mail, junichisuzuki-circ@umin.ac.jp

INTRODUCTION

Prostanoids are known to be multifunctional physiologic agents. They are produced by various stimuli and participate in the local homeostasis and physiologic responses. They are synthesized from arachidonic acid and are quickly released to the extracellular medium to send signals via prostanoid receptors [1]. Prostanoid receptors, DP, EP, FP, IP and TP are bound to PGD_2, PGE_2, $PGF_2\alpha$, PGI_2, and TXA_2, respectively [2]. In the cardiovascular system, these receptors express various cells, such as cardiomyocytes, vascular endothelium and smooth muscle cells [3]. They play an important role in modulating vascular homeostasis and one of their functions is participating in the development of vascular diseases such as thrombosis and atherosclerosis [4]. Recent studies revealed that prostanoids play an important role in both the occurrence and suppression of various cardiovascular diseases, such as myocardial ischemia, cardiac hypertrophy, cardiac fibrosis and atherosclerosis [5]. However, the precise role of prostanoids in cardiovascular disease has not yet been elucidated. To clarify the mechanisms of prostanoids, selective agonists/antagonists of the receptors and genetically modified mice have been developed.

Among the prostanoids, PGE_2 is known to play a key role in the pathogenesis of cardiovascular diseases. Wang *et al.* described that a deficiency of PGE_2 synthase-1 reduced plaque burden in fat-diet low-density lipoprotein receptor knockout mice [6]. Degousee *et al.* also reported that PGE_2 synthase-1 knockout mice showed impaired left ventricular contraction after acute myocardial infarction [7]. These studies indicate that PGE_2 plays a key role in inflammation and has pleiotropic effects [8, 9]. PGE_2 exerts their effects via G-protein–coupled receptors, and PGE_2-mediated cardioprotection involves one of the EP receptor subtypes, namely EP1, EP2 EP3, and EP4 [10]. Each receptor has their unique signaling pathways. EP1 couples to Gq and increases intracellular Ca2+. EP3 couples to Gi and inhibit the increase of cAMP. EP2 and EP4 receptors are coupled to stimulate adenylate cyclase, which leads to the elevation of intracellular cAMP [11]. In these EP receptors, recent studies showed EP3 and EP4 have cardioprotective effects in various *in vitro* and *in vivo* models. Recently, it was reported that PGE_2 suppresses chemokines such as monocyte chemoattractant protein (MCP)-1, macrophage inflammatory protein (MIP)-1α, MIP-1β and IFN-inducible protein-10 (IP-10) production in human macrophages via EP4 receptors [12]. The activation of EP4 also inhibits the production of proinflammatory cytokines such as TNF-α and IL-12 *in vitro* [13]. Among the receptors, EP2 and EP4 inhibited mixed lymphocytes reaction (MLR) and suppressed of ICAM-1, B7.1 and B7.2 production [14]. These studies suggest that EPs play a pivotal role in anti-inflammation.

In this review article, we will outline the recent progress of EPs and their role in inflammation-related cardiovascular diseases.

Myocardial Ischemia

In the development of myocardial infarction (MI) and ischemia reperfusion (I/R) injury, various inflammatory cells migrate into the heart and inflammatory cytokines and chemokines are produced. Cytokines including TNF-α and IL-1β evoke MCP-1 production from various cells, including inflammatory cells, cardiac myocytes and fibroblasts [15-17]. In this situation, it has been reported that PGE_2 shows cardioprotective effect with attenuating the activity of inflammatory cells, especially macrophages, and suppresses the production of various inflammatory cytokines and chemokines in the heart.

Although the exact mechanism of the cardioprotective role of EP3 is unclear, recent study showed that EP3 constitutively expressed on cardiomyocytes. It is functionally active and resulting in an inhibition of adenylate cyclase and subsequent decrease in cardiomyocyte cAMP formation. Thus, EP3 receptor overexpression on cardiomyocytes reduced ischemic myocardial injury, as indicated by an attenuation of contracture during ischemia and a reduction in the release of intracellular markers of ischemic injury during reperfusion [18]. Hohlfeld *et al.* demonstrated that the properties of EP3 include inhibition of adenylyl cyclase and upregulation during regional myocardial ischemia, suggesting an involvement in the anti-ischemic activity of PGE_2 and EP3 [19].

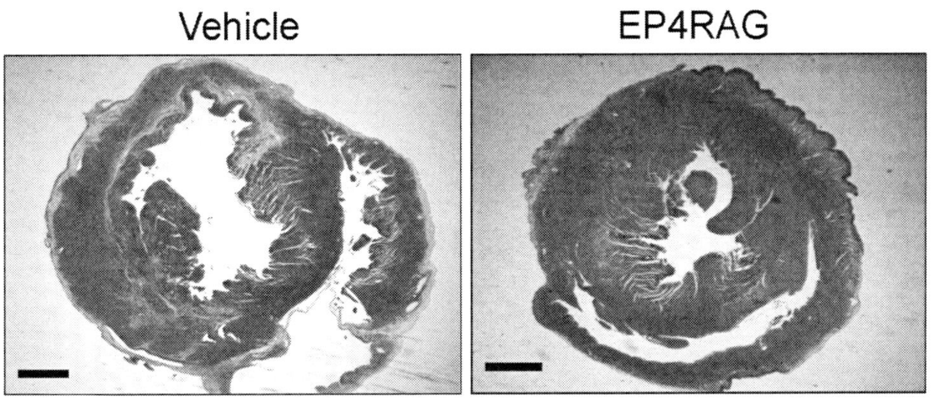

Although non-treated hearts showed left ventricular enlargement with expanded infarction and fibrosis, EP4RAG treatment significantly suppressed the changes.

Figure 1. Representative pathological findings of myocardial ischemia/reperfusion injury on day 7 (Masson staining, original magnification x 2)

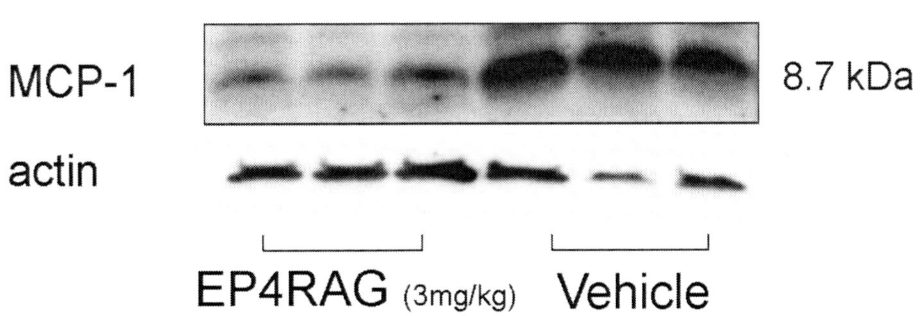

Although non-treated heart showed enhanced MCP-1 production, EP4RAG treatment significantly suppressed the levels.

Figure 2. Representative Western Blot findings of myocardial ischemia/reperfusion injury on day 7

To analyze the role of EP4 in the development of myocardial ischemia, we developed a novel EP4 selective agonist, which was named as EP4RAG. We administered EP4RAG or vehicle to rats with myocardial I/R injury. After 7 days of reperfusion, I/R rat hearts exhibited left ventricular dilatation and contractile dysfunction with myocyte hypertrophy and interstitial fibrosis (Figure 1). EP4RAG significantly reduced infarction area/ischemic myocardium and improved the ventricular contraction and dilatation compared to that of vehicle. EP4RAG also attenuated recruitment of inflammatory cells, especially macrophages and interstitial fibrosis in hearts. Western blot analysis and RNase protection assay showed that MCP-1 and other cytokines increased in both nonischemic and ischemic myocardium, however EP4RAG suppressed these changes (Figure 2). Gelatin zymography revealed that EP4RAG significantly reduced MMP-2 and -9 activities in both the nonischemic and ischemic area. Chemoattractant assay demonstrated EP4RAG suppressed the migration of cytokine-stimulated macrophages and decreased that the level of MCP-1 production *in vitro* [20].

Cardiac Transplantation

Cardiac rejection is a major complication in heart transplantation; cytokines, chemokines and adhesion molecules play critical roles in the development of rejection [21]. In the pathophysiology of rejection, expression of PGE_2 and EPs play a critical role in its progression. As Stone *et al.* demonstrated, administration of 16,16-dimethyl prostaglandin E2 (DMPGE2), a stable analogue of PGE_2, significantly prolonged the survival of heterotopic rat cardiac allografts [22]. Further, Kamei *et al.* revealed that a continuous infusion of DMPGE2 directly into cardiac allografts resulted in long-term graft survival and the development of recipient donor-specific tolerance [23]. Wanders *et al.* also investigated the influence of systemic injection of two compounds which inhibit the endogenous degradation of PGE_2 in a rat cardiac transplant model. They demonstrated that both compounds prolonged cardiac allograft survival [24]. Regarding the above, PGE_2 significantly prolonged heart allograft survival.

On the other hand, few papers reported the effects of EPs in cardiac transplantation. Although a paper demonstrated that EP2 and EP4 agonists prolonged cardiac allograft survival [25], there was little mention of the immunological mechanism between EPs and rejection in the report. Therefore, we used EP4RAG in a murine cardiac transplantation model to confirm the role of EPs in cardiac rejection [26]. In the paper, we revealed that the selective pharmacological activation of EP4 significantly prolonged graft survival compared to that of vehicle treatment (Figure 3). Pathologically, EP4RAG significantly suppressed myocardial cell infiltration with suppression of several proinflammatory factors such as cytokines and chemokines (Figure 4). *In vitro* study revealed that EP4RAG inhibited the activation of NF-κB through the suppression of macrophage activation, but it did not affect to T lymphocytes. However, effects of other EPs in cardiac transplantation have not yet been elucidated. Further investigation is needed to clarify the pathophysiology of prostanoids and rejection.

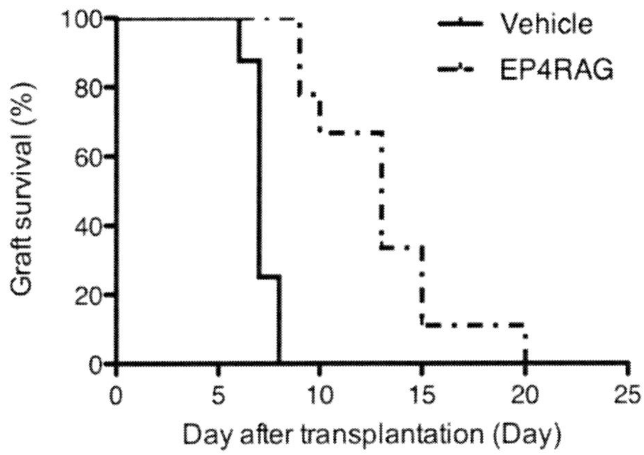

EP4RAG significantly prolonged graft survival compared to that of vehicle treatment.

Figure 3. Cardiac allograft survival

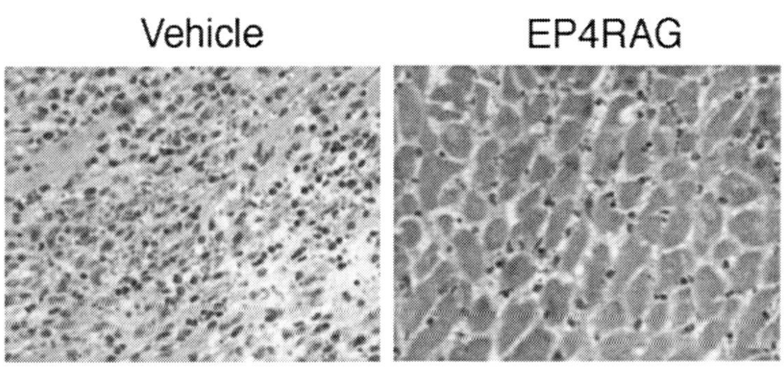

Although non-treated cardiac allografts showed significant myocardial cell infiltration, EP4RAG treatment significantly suppressed the changes.

Figures 1 and 2 are modified from reference 20, and figures 3 and 4 are modified from reference 26.

Figure 4. Representative pathological findings of allograft myocardium on day 7 (HE staining, original magnification x 400).

Conclusion and Future Direction

We showed that some members of prostanoids mediate the suppressive effects of myocardial inflammatory diseases. Especially, we showed a selective EP4 agonist attenuated inflammatory cytokines and chemokines via attenuation of macrophage activation in the cardiac I/R injury and allograft rejection. These findings of the PGE_2 and EPs would contribute to the development of new drugs targeting the prostanoid-induced inflammatory diseases.

REFERENCES

[1] Ukropec, J; Ukropcova, B; Kurdiova, T; et al: Adipose tissue and skeletal muscle plasticity modulates metabolic health. *Arch Physiol Biochem*, 2008, 114, 357-68.

[2] Narumiya, S; Sugimoto, Y; Ushikubi, F. Prostanoid receptors: structures, properties, and functions. *Physiol Rev*, 1999, 79, 1193-226.

[3] Cha, YI; Solnica-Krezel, L; DuBois, RN. Fishing for prostanoids: deciphering the developmental functions of cyclooxygenase-derived prostaglandins. *Dev Biol*, 2006, 289, 263-72.

[4] Jabbour, HN; Sales, KJ; Smith, OP; et al. Prostaglandin receptors are mediators of vascular function in endometrial pathologies. *Mol Cell Endocrinol*, 2006, 252, 191-200.

[5] Norel, X. Prostanoid receptors in the human vascular wall. *Scientific World Journal*, 2007, 7, 1359-74.

[6] Wang, M; Zukas, AM; Hui, Y; et al. Deletion of microsomal prostaglandin E synthase-1 augments prostacyclin and retards atherogenesis. *Proc Natl Acad Sci*, USA, 2006, 103, 14507-14512.

[7] Degousee, N; Fazel, S; Angoulvant, D; et al. Microsomal prostaglandin E2 synthase-1 deletion leads to adverse left ventricular remodeling after myocardial infarction. *Circulation*, 2008, 117, 1701- 1710.

[8] Tilley, SL; Coffman, TM; Koller, BH. Mixed messages: modulation of inflammation and immune responses by prostaglandins and thromboxanes. *J Clin Invest*, 2001, 108, 15-23.

[9] Betz, M; Fox, BS. Prostaglandin E2 inhibits production of Th1 lymphokines but not of Th2 lymphokines. *J Immunol*, 1991, 146, 108-13.

[10] Coleman, RA; Smith, WL; Narumiya, S. International Union of Pharmacology classification of prostanoid receptors: properties, distribution, and structure of the receptors and their subtypes. *Pharmacol Rev*, 1994, 46, 205-29.

[11] Southall, MD; Vasko, MR. Prostaglandin receptor subtypes, EP3C and EP4, mediate the prostaglandin E2-induced cAMP production and sensitization of sensory neurons. *J Biol Chem*, 2001, 276, 16083-91.

[12] Takayama, K; Garcia-Cardena, G; Sukhova, GK; et al. Prostaglandin E2 suppresses chemokine production in human macrophages through the EP4 receptor. *J Biol Chem*, 2002, 277, 44147-54.

[13] Nataraj, C; Thomas, DW; Tilley, SL; et al. Receptors for prostaglandin E(2) that regulate cellular immune responses in the mouse. *J Clin Invest*, 2001, 108, 1229-35.

[14] Morichika, T; Takahashi, HK; Iwagaki, H; et al. Effect of prostaglandin E2 on intercellular adhesion molecule-1 and B7 expression in mixed lymphocyte reaction. *Transplantation*, 2003, 75, 2100-5.

[15] Pfeffer MA; Braunwald E: Ventricular remodeling after myocardial infarction. Experimental observations and clinical implications. Circulation 1990, 81: 1161-72.

[16] Cohen, MV; Baines, CP; Downey, JM. Ischemic preconditioning: From adenosine receptor to KATP channel. *Ann Rev Physiol*, 2000, 62, 79-109.

[17] Bolli, R. The late phase of preconditioning. *Circ Res*, 2000, 87, 972-983.

[18] Martin, M; Meyer-Kirchrath, J; Kaber, G; et al. Cardiospecific overexpression of the prostaglandin EP3 receptor attenuates ischemia-induced myocardial injury. *Circulation*,

2005, 112, 400-6.

[19] Hohlfeld, T; Meyer-Kirchrath, J; Vogel, YC; et al. Reduction of infarct size by selective stimulation of prostaglandin EP(3) receptors in the reperfused ischemic pig heart. *J Mol Cell Cardiol*, 2000, 32, 285-96.

[20] Hishikari, K; Suzuki, J; Ogawa, M; et al. Pharmacological activation of the prostaglandin E2 receptor EP4 improves cardiac function after myocardial ischaemia/reperfusion injury. *Cardiovasc Res,* 2009, 81, 123-32.

[21] Taylor, DO; Edwards, LB; Aurora, P; et al. Registry of the International Society for Heart and Lung Transplantation: twenty-fifth official adult heart transplant report-2008. *J Heart Lung Transplant*, 2008, 27, 943-56.

[22] Stone, CD; Rosengard, BR; Boorstein, SM; et al. Prostaglandin E2 inhibits in vitro and in vivo lymphocyte responses in allogeneic transplantation. *Ann Thorac Surg*, 1990, 49, 927-30.

[23] Kamei, T; Callery, MP; Flye, MW. Intragraft delivery of 16, 16-dimethyl PGE2 induces donor-specific tolerance in rat cardiac allograft recipients. *Transplantation*, 1991, 51, 242-6.

[24] Wanders, A; Tufveson, G; Gerdin, B. Effects of prostaglandin E2 (PGE2) and drugs affecting PGE2 degradation on acute rejection of rat cardiac allografts. *Scand J Thorac Cardiovasc Surg*, 1992, 26, 33-7.

[25] Nomi, T; Sho, M; Akahori, T; et al. Protective effect of prostaglandin E2 receptors EP2 and EP4 in alloimmune response in vivo. *Transplant Proc*, 2006, 38, 3209-10.

[26] Ogawa, M; Suzuki, J; Kosuge, H; et al. The mechanism of anti-inflammatory effects of prostaglandin E_2 receptor 4 activation in murine cardiac transplantation. *Transplantation*, 2009, 87, 1645-53.

In: Prostaglandins: Biochemistry, Functions… ISBN: 978-1-61668-272-9
Editors: Gillian M. Goodwin, pp.135-165

Chapter 10

REGULATION OF PROINFLAMMATORY TRANSCRIPTION FACTORS BY CYCLOPENTENONE PROSTAGLANDINS: INVOLVEMENT OF DIRECT AND INDIRECT MECHANISMS*

Eva Cernuda-Morollón and Dolores Pérez-Sala †
Departamento de Estructura y Función de Proteínas. Centro de Investigaciones Biológicas. Consejo Superior de Investigaciones Científicas (C.S.I.C.), Madrid, Spain

ABSTRACT

Inflammatory responses are essential defense mechanisms which protect pluricellular organisms from various types of insults and imply the regulation of a vast number of genes. The transcription factors NF-κB and AP-1 are key players in inflammation, since they are rapidly activated by inflammatory stimuli and in turn direct the expression of inflammatory mediators such as cytokines, matrix metalloproteinases, iNOS and COX-2. The inflammatory response is a tightly controlled process. It possesses intrinsic mechanisms that limit its intensity and duration. Key inflammatory mediators such as nitric oxide and prostaglandins may contribute both to the onset and to the resolution of inflammation. Cyclopentenone prostaglandins (cyPG) are reactive compounds which may form covalent adducts with thiol groups in proteins or in glutathione. CyPG have proven potent antiinflammatory agents both in cellular and in animal models. CyPG may act as ligands for the nuclear receptors of the PPAR family of transcription factors, and PPAR-mediated responses appear to contribute to their beneficial effects. In addition, posttranslational protein modification by covalent cyPG binding may modulate signal transduction cascades leading to transcription factor regulation and/or affect the activity of transcription factors themselves. The transcription factors NF-κB and AP-1 have been identified as targets for this modification. In this work we have explored the effects of

* A version of this chapter was also published in *Trends in DNA Research*, edited by Corey R. Woods published by Nova Science Publishers, Inc. It was submitted for appropriate modifications in an effort to encourage wider dissemination of research

† Corresponding author: Phone: 34 918373112; Fax: 34 915360432; email: dperezsala@cib.csic.es

cyPG on the NF-κB activation pathway in mesangial cells, a known cellular model of inflammation. Activation of mesangial cells in the presence of the cyPG 15-deoxy-$\Delta^{12,14}$-prostaglandin J_2 reduces NF-κB transcriptional activity. Pre-treatment with cyPG inhibits NF-κB DNA binding activity both in intact cells and in isolated nuclear extracts. By using a biotinylated 15-deoxy-$\Delta^{12,14}$-prostaglandin J_2 derivative we have observed the modification of the NF-κB subunits p65 and p50 in intact cells. Subsequently, we have used biotinylated cyPG to study the interaction of these prostanoids with NF-κB in vitro. Binding of biotinylated cyPG to p50 is selectively reduced in the presence of an oligonucleotide containing the consensus site for NF-κB binding, which suggests that this interaction involves the DNA binding domain. Labeling studies of a cys62ser p50 mutant confirm that this cysteine residue is a site for cyPG attachment. Biotinylated cyPG labeling assays indicate that other proinflammatory transcription factors are potential targets for modification by cyPG. In addition, cyPG may modulate the activity of proteins along the activation pathways of these factors. Moreover, cyPG may alter cellular redox state by various mechanisms. Given the fact that numerous transcription factors, including NF-κB and AP-1 are redox-regulated, both direct and indirect mechanisms can be envisaged for the modulation of pro-inflammatory transcription factors by cyPG. Finally, a number of endogenous lipids as well as natural products possess reactive unsaturated carbonyl groups, the approaches used in this study could prove useful to assess their potential in the regulation of transcription factor activity.

INTRODUCTION

The inflammatory response is a complex process by which organisms react to trauma or infection. It implies the interaction between multiple soluble factors and many types of cells in a tightly controlled way [1]. Tissue injury induces the release of bioactive peptides from neurons, and of intracellular products from damaged cells. Immune cells present in tissues and leukocytes recruited from the blood vessels release cytokines, eicosanoids and histamine that contribute to vasodilatation and edema [1]. Consequently other mediators are activated, including matrix metalloproteinases and the systems involved in the generation of reactive oxygen and nitrogen species, such as superoxide anion and nitric oxide (NO). The aim of this rapid response is to kill invasive bacteria and destroy damaged tissue, but for this process to lead to healing it is necessary that negative control mechanisms are in place to avoid excessive damage or the development of a chronic situation. Several inflammatory mediators display a dual role that can either promote or suppress inflammation depending on their concentration or the activation state of the target cells.

NO and prostaglandins (PG) are key mediators of the inflammatory process. In inflammatory settings the main source of NO is the inducible form of nitric oxide synthase or iNOS. iNOS is not normally expressed in most tissues but its levels increase rapidly in response to proinflammatory agents such as cytokines or bacterial lipopolysaccharide (LPS). iNOS expression is tightly controlled by transcriptional and posttranscriptional mechanisms [2]. iNOS can generate high amounts of NO, a free radical which may directly modulate protein function by modifying specific residues. NO can also regulate the activity of enzymes such as cyclooxygenases and soluble guanylate cyclase through the interaction with the heme group of these enzymes. The activation of soluble guanylate cyclase leads to an increased generation of cGMP, one of the main mediators of NO action. NO plays a multifunctional role in inflammation [3]. Among the proinflammatory actions of NO are the induction of vasodilation and the stimulation of the expression of proinflammatory genes. It has been observed that NO can also contribute to the resolution of inflammation through the induction of macrophage apoptosis, and the regulation of gene expression [4].

PG released in inflammatory situations contribute to vasodilatation, edema and fever. Therefore PG mediate many of the symptoms of acute and chronic inflammation. PG are synthesized from arachidonic acid by several enzymatic steps. Arachidonic acid is released from membrane phospholipids by the action of phospholipases. The induction of cyclooxygenase-2 expression, the rate limiting enzyme in the transformation of arachidonic acid into PG, plays a key role in the increase of PG associated with inflammatory situations. COX-2 catalyzes the formation of PGH_2, which is in turn converted by isomerases into the different PG in a tissue- or cell type-related fashion. In addition to enzymatic synthesis, several types of PG can arise from the non-enzymatic transformation of arachidonic acid or of other prostaglandins.

The interaction between iNOS and COX-2 pathways plays a central role in the regulation of inflammation. The expression of iNOS and COX-2 appears to be coordinately regulated in many settings. Numerous proinflammatory stimuli trigger the expression of both enzymes. In addition, this induction can be subjected to the modulatory action of antiinflammatory cytokines, enzyme cofactors, corticosteroids and reactive oxygen species, which influence the levels of both proteins in a similar way [5-7]. We and others have explored the mechanisms underlying the interplay between iNOS and COX-2 induction in a cellular model of inflammation such as renal glomerular mesangial cells (MC) exposed to proinflammatory stimuli [8-11]. MC are specialized smooth muscle cells which are located around the capillaries of the renal glomerulus where they play an important role in the maintenance of glomerular structure and regulation of glomerular filtration rate [12, 13]. Upon exposure to inflammatory stimuli, these cells suffer a trans-differentiation process to an activated phenotype and become capable of releasing inflammatory mediators such as NO, eicosanoids, chemokines and cytokines. Activated MC also express integrins and adhesion molecules which participate in the recruitment and activation of macrophages. MC express iNOS and COX-2 in response to proinflammatory agents, which are responsible for an increased generation of NO and prostaglandins [14, 15].

When analyzing the cross-talk between the NOS and COX pathways it is important to consider that the products of these enzymes may exert a feedback regulation of both pathways, as shown schematically in Figure 1. NO has been reported to modulate COX activity both *in vitro* and *in vivo* [16-18]. NO may also modulate COX-2 expression. The effects of NO on the levels of COX-2 are complex and depend on the cell type, concentration, and time. Both amplifying and inhibitory effects of NO on COX-2 expression have been reported in diverse experimental settings [19-21]. We have previously described that NO donors can exert a biphasic regulation of COX-2 levels in mesangial cells [11]. NO also modulates iNOS levels through multiple mechanisms. This modulation can also follow a biphasic pattern [22]. We have postulated that this type of effect may contribute to limit the amplitude or the duration of the inflammatory response. Interestingly, COX products may modulate both iNOS and COX-2 expression. In the early stages of inflammation, COX products may contribute to an amplification of the induction of these proteins, while at later instances or when they reach higher levels they can display an inhibitory effect [23]. A switch from a proinflammatory to an antiinflammatory nature and/or role of the eicosanoids generated during the inflammatory process has been reported that will contribute to the onset and the resolution of inflammation [24, 25].

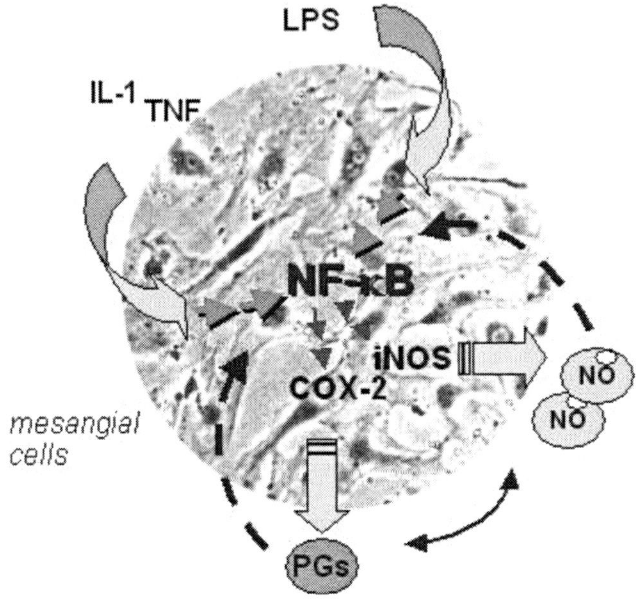

Figure 1. The interdependence of iNOS, COX-2 and NF-κB signaling pathways plays a key role in the regulation of the inflammatory response in mesangial cells. Proinflammatory stimuli trigger the expression of iNOS and COX-2 in mesangial cells. The NO and PG biosynthetic pathways are subjected to significant cross-talk since their products may regulate the activity and/or the expression of iNOS and COX-2. The transcription factor NF-κB mediates the induction of iNOS and COX-2 by cytokines and is an important target for the regulatory effects of NO and PGs on gene expression.

The molecular mechanisms by which this intricate regulation takes place occur at multiple levels, including the interaction with membrane receptors or intracellular proteins, the generation of second messengers of NO such as cGMP and of PG like cAMP. An important level of regulation involves the modulation of transcription factor activity. Many transcription factors contribute to the inflammatory response. Among the most extensively studied are AP-1 and NF-κB. The transcription factor AP-1 is formed by homo or heterodimers of proteins of the Jun and Fos families. AP-1 is activated in response to cytokines, mitogenic signals and cellular stress [26]. Regulatory sites for the binding of AP-1 are present in the promoters of numerous proinflammatory genes, including iNOS, COX-2 and matrix metalloproteinases. The transcription factor NF-κB is considered a key player in inflammation. NF-κB is a dimeric factor formed by members of the Rel family of proteins [27]. The prototypical member of the NF-κB family is a heterodimer consisting of p65/RelA and p50/NF-κB1. Both subunits possess DNA binding activity, but only the p65 subunit displays transactivating activity. In a very simplified view, stimuli such as cytokines and mitogenic signals initiate a cascade of events that leads to the activation of a kinase, known as IKK, which specifically phosphorylates IκB, the inhibitory subunit of NF-κB [28]. This triggers the ubiquitination and degradation of IκB by the proteasome and favors the accumulation of the active NF-κB dimers in the cell nucleus where the factor binds to sites in the promoters of many genes involved in inflammation such as iNOS, COX-2, intercellular adhesion molecule-1 (ICAM-1) and monocyte chemoattractant protein-1 (MCP-1), among others. Mutation of these sites frequently results in a drastic reduction in promoter inducibility by proinflammatory stimuli [7]. As schematized in Figure 1, NO and PG exert an important feedback regulation of NF-κB. NO can bring about both positive and negative effects on NF-κB activity. NO can activate NF-κB by a Ras-mediated mechanism depending on the direct modification of this signaling protein by S-nitrosylation at cysteine 118. A NO-dependent amplification of NF-κB activation by proinflammatory stimuli has also been observed, which has been shown to correlate with a delay in the resynthesis of IκB [11]. The generation of elevated levels of NO when the inflammatory response is established may exert negative effects on NF-κB activation, and this in turn may constitute an important mechanism contributing to the resolution of inflammation. In fact, the effect of genetic or pharmacological suppression of iNOS may lead to biphasic effects on the inflammatory response [22, 29, 30]. The inhibition of NF-κB by NO can occur by cGMP-dependent and independent mechanisms. NO can amplify IκB expression and interfere with the phosphorylation and degradation of IκB in a cGMP-independent manner [31, 32]. Recently, a mechanism for the direct inhibition of NF-κB DNA binding by NO has been characterized. NO can directly S-nitrosylate a cysteine residue located in the DNA binding domain of the NF-κB subunit p50, cysteine 62 in human p50, which needs to be in a reduced state to allow efficient DNA binding [33]. Moreover, in the presence of physiological concentrations of glutathione, NO donors can induce the formation of a mixed disulfide between glutathione and p50 which involves cysteine 62 [34]. In addition, NO may induce other oxidative modifications of NF-κB proteins that may contribute to the inhibitory effect [34].

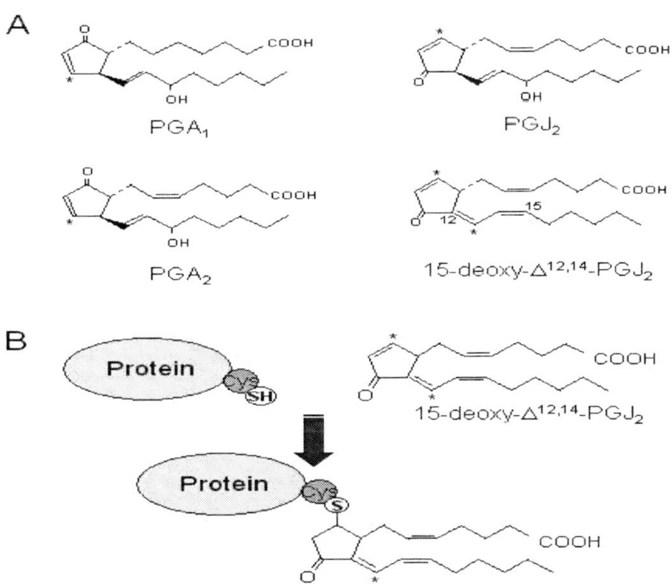

Figure 2. Structure of cyPG and formation of Michael adducts with proteins. (A) Structures of several cyPG of the A series (left) and of the J series (right). Electrophilic carbons are marked by asterisks. (B) Schematic representation of the formation of a Michael adduct between the electrophilic carbon in the cyclopentenone ring of 15d-PGJ_2 and the sulphydryl group of a protein cysteine residue.

NF-κB activity is also regulated by prostanoids. PG interacting with membrane receptors can lead to an increase in cAMP concentrations. cAMP in turn can modulate NF-κB activation by several mechanisms, including the phosphorylation of IκB and of the NF-κB subunit p65 by cAMP-dependent protein kinase, which leads to increased activation, which can be dependent in part on the recruitment of coactivators [35, 36]. A direct increase in the transactivation potential of p65 without changes in nuclear accumulation has been invoked for the potentiation of TNF-elicited NF-κB activation by PGE or PGE analogs [37]. In addition, increases in cAMP triggered by treatment with PGE have been shown to correlate with an induction of the mRNA of p65 [38]. In some experimental settings, PGE_2 can inhibit NF-κB activation [39]. The nature of the modulation of the inflammatory response by PGE_2 may thus depend on the cell type or the nature of the co-stimulatory signal. Recently, much attention has been devoted to a type of PG known as cyclopentenone prostaglandins (cyPG) (Figure 2A).

These endogenous eicosanoids are generated by the spontaneous dehydration of their parent PG. More precisely, PG of the A series arise from the dehydration of PGE, while PG of the J series are produced by dehydration of PGD. These PG owe their name to the presence in their structure of an α,β-unsaturated carbonyl group in the cyclopentane ring (cyclopentenone). CyPG have been detected in biological fluids and tissues in significant amounts. Although the magnitude of the generation and the bioavailability of these compounds is still a matter of controversy, increases in cyPG generation have been detected in association with COX-2 induction in several pathophysiological situations. The cyPG 15-deoxy-$\Delta^{12,14}$-PGJ_2 (15d-PGJ_2) has been detected in the foamy macrophages of human atherosclerotic plaques [40] and in spinal cord sections of patients with sporadic amyotrophic lateral sclerosis [41]. CyPG, and in particular 15d-PGJ_2 have been detected also in experimental cellular and animal models of inflammation. The exploration of the biological effects of these eicosanoids has led to the observation of their antitumoral, antiviral and antiinflammatory effects. The antiviral effects of cyPG appear to be associated with the induction of a heat shock response in cells, the induction of IκBα expression and the inhibition of NF-κB activity [42, 43]. Recently, 15d-PGJ_2 has been identified as a potential endogenous ligand of the transcription factor known as PPARγ (peroxisome proliferator activated receptor γ) [44, 45]. The PPAR subfamily of nuclear receptors is constituted by three main forms, known as α, β and γ, which differ in agonist selectivity and tissue distribution. PPAR are ligand activated transcription factors involved in the modulation of lipid metabolism and inflammation [46, 47]. Since PPAR activation was found to counteract inflammation in many experimental models, this provided a mechanism for the antiinflammatory effects of cyPG. Based on the detection and the effects of 15d-PGJ_2 at the late stages of inflammation in a rat model of pleurisy, an antiinflammatory role for COX-2 was proposed [24]. During the last several years, negative effects on cyPG on proinflammatory gene expression have been frequently reported. CyPG have been shown to inhibit the induction of numerous genes involved in the pathophysiology of inflammation, including iNOS, COX-2, ICAM-1, MCP-1 and matrix metalloproteinase-9 (MMP-9) [48, 49]. For these reasons the potential for the pharmacological application of these or related agents as antiinflammatory drugs is being pursued.

A common mechanism underlying the antiinflammatory effects of cyPG in the various experimental systems is the inhibition of NF-κB. This effect can be due to several mechanisms. As stated above, some cyPG including 15d-PGJ_2 and PGA_1 can bind and activate PPAR. These transcription factors, in turn, may interfere with NF-κB activation by competing for limiting amounts of coactivators, by modulating the expression of IκB or by direct protein-protein interaction [50-52]. A PPAR-dependent effect of 15d-PGJ_2 has been reported to be responsible for the inhibition of iNOS and COX-2 expression in cardiomyocytes [53], and for the inhibition of iNOS and MMP-13 in human chondrocytes [54]. However, cyPG can also display antiinflammatory actions by PPAR-independent mechanisms. CyPG can induce oxidative stress and this in turn may be related to the induction of apoptosis in macrophages, a phenomenon that has been proposed to contribute to the resolution of inflammation [55], or mediate the induction of genes with antiinflammatory potential, such as heme-oxygenase [56].

The cyclopentenone moiety is an important structural feature of cyPG. This confers a strong reactivity to the carbon at the –β position in the cyclopentenone ring (Figure 2) which may form adducts with cellular thiols, such as the sulphydryl group of glutathione or of protein cysteine residues [57], by a reaction known as Michael addition, which appears schematized in Figure 2B. Some cyPG possess a dienone structure with two electrophilic carbons (marked in Figure 2 by asterisks), both of which can be involved in the formation of Michael adducts [58, 59]. The reaction of cyPG with cysteine residues in proteins forming stable complexes can have important consequences for protein function [60]. The importance of this process for the antiinflammatory actions of cyPG is becoming more clear as the potential targets for this type of modification, which could be termed prostanylation or eicosanylation, are identified. We and others have described that cyPG may directly interfere with transcription factors involved in the control of inflammation, including NF-κB and AP-1 [59, 61, 62]. This effect may involve the interaction with proteins forming the signaling cascade that leads to transcription factor activation, or the interaction with transcription factors themselves. In this chapter we have used a fragment from the human p50 subunit of NF-κB, which corresponds to the DNA binding domain, as a model to explore the direct interaction of cyPG with transcription factors. Moreover, we have explored the effects of 15d-PGJ_2 on NF-κB signaling in a cellular model of inflammation, such as mesangial cells (MC).

MATERIALS AND METHODS

Materials

15-Deoxy-$\Delta^{12,14}$-PGJ_2 was from Calbiochem-Novabiochem (San Diego, CA). 15-Deoxy-$\Delta^{12,14}$-PGJ_2 (mixture of isomers), PGE_2, PGA_1 and PGA_1-biotinimide were from Cayman Chemical (Ann Arbor, MI). Recombinant human IL-1β (5×10^7 U/mg) was from Roche (Germany). Recombinant human TNF-α was from Serotec (Oxford, U.K.). Rabbit antisera against human p50 and p65 proteins were a generous gift of Dr. Nancy Rice (National Cancer Institute, Bethesda, MD, USA). Horseradish peroxidase (HRP) conjugated anti-rabbit immunoglobulins were from Dako (Glostrup, Denmark). Polyclonal anti-IκBα (sc-371) was from Santa Cruz Biotechnology (Santa Cruz, CA). The recombinant wild-type and mutant DNA binding domains of human p50 (amino acids 36 - 385, GenBank accession number

M55643), expressed *in Escherichia coli* as hexahistidine fusion proteins and purified as described [34], were the generous gift of Dr. E. Pineda-Molina and Dr. S. Lamas. This construct will be referred to as p50. GSH was from Sigma. Ethylcarbodiimide, EZ-link 5-(biotinamido)pentylamine and biotinylated BSA standard were from Pierce (Rockford, IL). HRP-conjugated streptavidin was from Amersham Pharmacia Biotech (Buckinghamshire, U.K.). Cell culture media and supplements were from GIBCO Life Technologies S.A. (Barcelona, Spain).

Preparation of Biotinylated 15d-PGJ$_2$

Biotinylated 15d-PGJ$_2$ was prepared by a procedure similar to that previously used for the preparation of biotinylated PGA$_2$ [63]. Briefly, 1 mg of 15d-PGJ$_2$ was made to react with 1 mg of EZ-link (5-biotinamido)pentylamine in the presence of the condensing agent diisopropylcarbodiimide. The reaction was carried out in acetonitrile for 16 h at room temperature. Biotinylated 15d-PGJ$_2$ was extracted with ethyl acetate and purified through a silica gel column eluted with ethyl acetate:methanol (9:1). The concentration of biotinylated PG was estimated from the determination of biotin as previously described [62]. The modified PG was then dried under argon and dissolved in DMSO for further use.

Cells and Treatments

For this study, rat mesangial cells (RMC) between passages 10 and 15 were used. Primary cultures of RMC were obtained using a technique of graded sieving as reported earlier [15]. Groups of cells with mesangial morphology were subcloned and expanded. Mesangial cells displayed positive staining with anti-vimentin, anti-Thy 1.1, and anti-myosin, and absence of staining with anti-factor VIII [5]. Cells were grown in RPMI 1640 supplemented with 10% fetal bovine serum (FBS), 2 mM glutamine, 100 U/ml penicillin and 100 μg/ml streptomycin. For treatments, cells were incubated in RPMI medium without phenol red in the absence of serum. Cell activation was achieved by incubation with a combination of IL-1β (2 ng/ml) plus TNF-α (25 ng/ml) for the indicated time periods. Prostaglandins were added in DMSO to the incubation medium 15 min prior to cytokine addition. Final DMSO concentration was 0.1% (vol/vol).

Analysis of NF-κB Transcriptional Activity

NF-κB transcriptional activity was estimated in transient transfection experiments. MC at 80% confluence were transfected with 1.5 μg of a luciferase reporter plasmid containing three consensus NF-κB response elements (3xNF-κB-pConA-Luc) [64] or empty vector (pConA-Luc), plus 2 ng of the Renilla luciferase reporter vector pRL-CMV (Promega, Madison, WI) for 3 h in Optimem (Gibco) using the LIPOFECTAMINE™ Reagent (Invitrogen-Life technologies, Barcelona, Spain). After a 24 h incubation in serum-free medium, cells were treated with 15d-PGJ$_2$ for 15 min before stimulation with cytokines, as indicated. Firefly and

Renilla luminiscences were measured in cell lysates according to the instructions of the manufacturer using a dual luciferase reporter system from Promega. Promoter activities were expressed in arbitrary units as the ratio between the two luciferase measurements.

Preparation of Nuclear and Cytosolic Extracts

Nuclear extracts from RMC were obtained as previously described [11]. RMC were resuspended in 10 mM Hepes, pH 7.9, 10 mM KCl, 0.1 mM EDTA, 0.1 mM EGTA, 1 mM dithiotreitol (DTT), 0.1 mM sodium vanadate, 50 mM sodium fluoride, 0.5 mM PMSF, and 2 μg/ml of leupeptin, antipain, trypsin inhibitor and pepstatin A. NP-40 was added to a final concentration of 0.5%. After vortexing for 10 s, nuclei were sedimented by centrifugation for 5 min at 15000 rpm. The supernatants were collected and centrifuged again at 200000 x g for 30 min to obtain the cytosolic fraction. The nuclear pellets were extracted by shaking on ice for 30 min in 20 mM Hepes, pH 7.9, 400 mM NaCl, 1 mM EDTA, 1 mM EGTA, 1 mM DTT, 0.1 mM sodium vanadate, 50 mM sodium fluoride plus protease inhibitors as above. Nuclear debris was discarded by centrifugation at 15000 rpm for 5 min, and nuclear extracts were stored at –80°C. Protein content was determined using the BCA protein assay from Pierce (Rockford, IL) for cytosolic extracts, and the Bradford method for nuclear extracts.

Electrophoretic Mobility Shift Assays (EMSAs)

Sense and antisense oligonucleotides corresponding to the upstream NF-κB site from the human COX-2 promoter (5′-GGAGAGGGGATTCCCTGCG-3′), located within – 452 to – 433 bases from the transcriptional start site in the reported sequence [65] were annealed and labeled by incubation with [α-^{32}P] dCTP using the klenow fragment, and isolated by gel filtration on Sephadex G-50. For the binding reaction, 2 ng of the labeled oligonucleotide (approximately 20000 cpm) and 1 μg of poly(dI-dC) carrier was incubated with 3.5 μg of nuclear extract in binding buffer (4% glycerol, 1 mM $MgCl_2$, 0.5 mM EDTA, 0.5 mM DTT, 50 mM NaCl, 10 mM Tris-HCl, pH 7.5) for 20 min on ice. Protein-DNA complexes were separated by electrophoresis on a non-denaturing 6% polyacrylamide gel and visualized by autoradiography. To ensure the specificity of the complexes formed, competition experiments were carried out by performing the binding assays in the presence of a 200-fold excess of unlabeled NF-κB oligonucleotide or of an unrelated oligonucleotide corresponding to bases – 456 to – 432 from the transcriptional start site in the reported sequence of the bovine endothelial NOS gene [66].

Preparation of Cell Lysates and Western Blot

For analysis of protein levels in total cell lysates, cells were lysed in 10 mM Tris pH 7.5, 0.1 mM EDTA, 0.1 mM EGTA, 0.1 mM β-mercaptoethanol, 0.5% SDS, and 2 μg/ml of leupeptin, antipain, trypsin inhibitor and pepstatin A, by forced passes through a narrow needle on ice. Protein concentration was determined by the BCA protein assay from Pierce

(Rockford, IL). Fifteen μg of protein from each experimental condition were electrophoresed on 10% polyacrylamide gels and transferred to Immobilon-P membranes (Millipore, Bedford, MA). As a control for even protein loading and transfer, membranes were stained with Ponceau S. Blots were probed with primary antibodies anti-IκBα (at 1:1000 dilution) or anti-p65 and anti-p50 (at 1:2000 dilution) followed by HRP-conjugated secondary antibodies at 1:2000 dilution, and immunocomplexes were visualized using an enhanced chemiluminiscence (ECL) detection system from Amersham. The intensity of the bands was estimated by image scanning and corrected by the values obtained from the Coomassie staining of membranes or by the signal given by an antibody against a non-related protein.

Analysis of the NF-κB Dimers Present in Nuclear Extracts by Western Blot

Six μg of nuclear extracts were incubated with 400 ng of non-labeled NF-κB oligonucleotide, and the complexes formed were separated by gel electrophoresis, as above. Gels were transferred to Hybond-PVDF membranes and blots were probed with anti-p50 and anti-p65 antibodies before detection with ECL.

Biotinylated 15d-PGJ$_2$ Labeling of p50 in RMC

Pre-confluent RMC were incubated with 10 μM biotinylated 15d-PGJ$_2$ for 2 h. Cells were washed twice with PBS, harvested and lysed in 10 mM Tris-HCl pH 7.5, 0.1 mM EDTA, 0.1 mM EGTA, 1% NP-40, 0.1 mM β-mercaptoethanol plus protease inhibitors. Lysates were centrifuged for 10 min at 10000 x g to discard insoluble debris. Supernatants containing 200 μg of protein were incubated batch-wise with 50 μl of Soft-Link Avidin (Promega, Madison, WI) for 1 h at 4°C. Beads were washed 5 times with lysis buffer by centrifugation at 6000 rpm for 1 min. Proteins retained on the beads were eluted by boiling the beads in Laemmli sample buffer. The presence of p50, or other proteins of interest, in the eluate was assessed by Western blot.

Binding of Biotinylated cyPG to p50 in Vitro

The formation of covalent adducts between p50 and biotinylated cyPG, the structure of which is shown in Figure 3, was monitored by means of a Western blot-based assay. p50 in 20 mM Tris pH 7.5, 50 mM NaCl, 5mM $MgCl_2$, 1 mM EDTA, 0.7 μM β-mercaptoethanol, 0.1 mM DTT, 5% glycerol, 0.01% Nonidet P-40, was incubated with prostaglandins or vehicle (DMSO) at r.t. At the indicated time points Laemmli sample buffer was added and samples were analyzed by SDS-PAGE on 10% polyacrylamide gels. Proteins on gels were transferred to immobilon P membranes (Millipore) following the instructions of the manufacturer by using a semi-dry transfer blot apparatus from Bio-Rad. After blocking for 1 h by incubation with 2% non-fat dried milk in TBS-T (10 mM Tris-HCl, pH 7.5, 500 mM NaCl, 0.05% Tween 20), membranes were incubated with HRP-conjugated streptavidin

(1:1000 dilution in 1% BSA in TBS-T). Signal was developed with the ECL system from Amersham. Blots were stripped by brief incubation with 8 N guanidinium hydrochloride, pH 3.0, and incubated with anti-p50 antibodies at 1:2000 dilution. Under the conditions used in the assays the incorporation of biotin was linear with respect to time, protein amount and biotinylated cyPG concentration. Estimation of biotin incorporation was achieved by comparison of the signal with that given by a biotinylated BSA standard (Pierce). After 4 h incubation of 5 μM p50 in the presence of 10 μM biotinylated 15d-PGJ_2 at room temperature, approximately 0.1 nmoles of biotin were incorporated per nmole of protein.

RESULTS

15d-PGJ_2 Inhibits NF-κB Transcriptional Activity in RMC

Given the importance of the transcription factor NF-κB in the regulation of the induction of numerous proinflammatory genes, we were interested in studying the consequences of the interaction of cyPG with the NF-κB pathway in a model of inflammation such as RMC. For this purpose we first studied the effect of 15d-PGJ_2 on NF-κB transcriptional activity. RMC were transiently transfected with 1.5 μg of a NF-κB reporter plasmid containing 3 copies of NF-κB binding sequence driving the expression of luciferase (3xNF-κB-conA-luc) and 2 ng of pRL-CMV, as described in Methods. After a recovery period of 24 hours, RMC were treated with vehicle or 15d-PGJ_2 and 15 min later they were stimulated with IL-1β plus TNF-α (Ck) for 6 hours. The transcriptional activity of NF-κB was estimated from the luciferase activity present in cell lysates. Under these conditions the PG did not affect the basal activity of the reporter but inhibited cytokine-elicited induction of NF-κB activity by 30% (Figure 4).

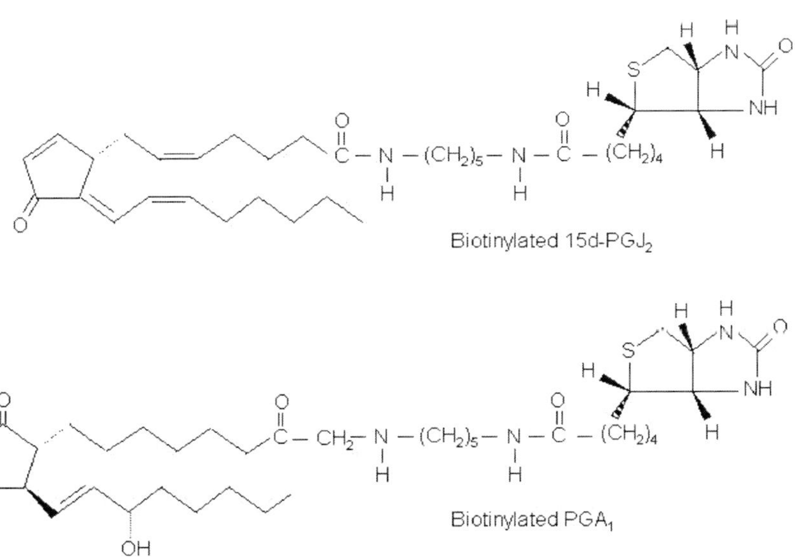

Figure 3. Structures of the biotinylated derivatives of 15d-PGJ_2 and PGA_1 used in this study.

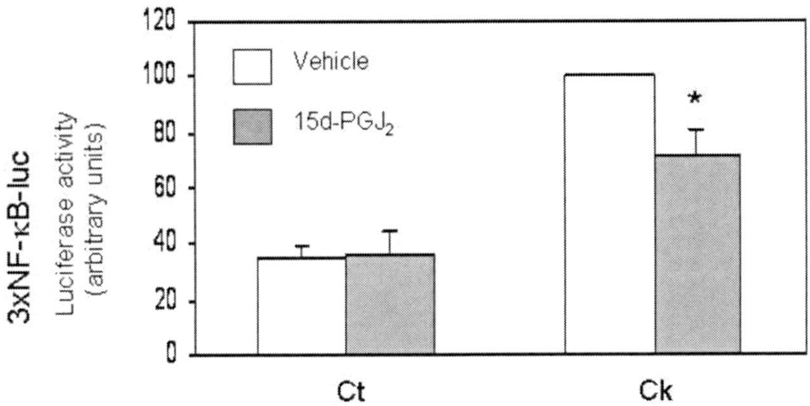

Figure 4. Effect of 15d-PGJ_2 on NF-κB transcriptional activity in RMC. The transcriptional activity of NF-κB in RMC was evaluated by transient transfection of a 3xNF-κB reporter construct. Cells were treated with 2 ng/ml IL-1β plus 25 ng/ml TNF-α (Ck) in presence or absence of 5 μM 15d-PGJ_2 for 6 hours. Results are average values ± SEM of three experiments performed in triplicate. (*$p<0.05$ *versus* vehicle).

Effect of 15d-PGJ_2 on NF-κB DNA Binding Activity in Cytokine-stimulated RMC

We were then interested in exploring which steps in the NF-κB activation pathway were affected by 15d-PGJ_2 treatment. We first assessed whether 15d-PGJ_2 inhibited NF-κB DNA binding activity in intact cells. For this purpose we performed EMSA assays with nuclear extracts obtained from cells activated with Ck in presence or absence of the PG. Confluent RMC, kept in serum-free medium 24 hours before the experiment, were treated with the indicated agents at different time points. Activation of RMC with the cytokine mixture resulted in an induction in the formation of the NF-κB complex (Figure 5A), as previously described by us [11]. The specificity of this complex was assessed by competition assays performed in the presence of excess unlabeled NF-κB oligonucleotide (comp), which completely abolished binding of the radiolabeled probe, or of an unrelated oligonucleotide (unrelated), which did not block complex formation. In nuclear extracts from cells activated in the presence of 15d-PGJ_2 the NF-κB-DNA binding activity was clearly reduced. The inhibition of DNA-protein complex formation was apparent at all time points studied.

To gain insight into the mechanism of the inhibition of NF-κB-DNA binding we analyzed the levels of p50 and p65 subunits in the nuclear extracts by Western blot (Figure 5B). We observed that after 5 minutes of treatment with cytokines, there were not detectable differences in the nuclear levels of p65 and p50 in cells activated in the absence or presence of 15d-PGJ_2. However, at this time point, there was an intense inhibition of NF-κB-DNA complex formation (43%) in 15d-PGJ_2-treated cells.

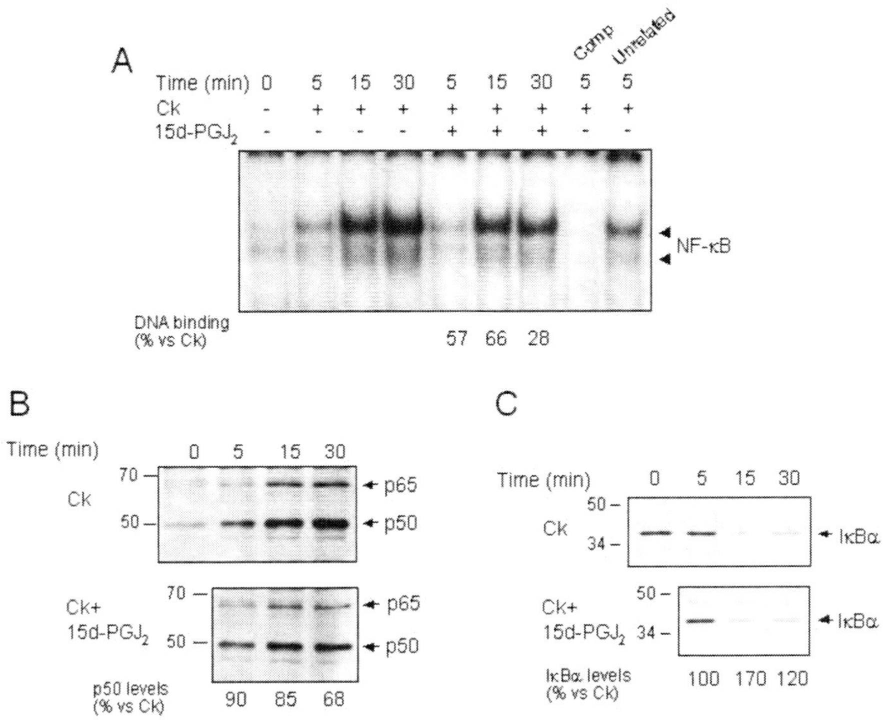

Figure 5. Effect of 15d-PGJ$_2$ on DNA binding and levels of NF-κB proteins. A) RMC were activated with Ck in presence or absence of 5 μM 15d-PGJ$_2$ and nuclear extracts were isolated at the indicated time points. The NF-κB-DNA binding activity was analyzed by EMSA. The formation of NF-κB-DNA complexes was quantitated and it is expressed as percentage of the values obtained in the absence of 15d-PGJ$_2$. Results shown are representative of two experiments. Levels of nuclear p50 and p65 (B) and cytosolic IκBα (C) were estimated by Western blot. The position of molecular mass standards (KDa) is shown on the left of the panels. Results are representative of three independent experiments.

This suggests that 15d-PGJ$_2$ may directly impair DNA binding in this model system. However, after 30 minutes of stimulation, there was a moderate reduction in the nuclear levels of p65 and p50 in 15d-PGJ$_2$-treated cells (32 % of the amount detected in cells treated with cytokines alone at the same time point). This suggests that a partial inhibition of NF-κB translocation to the nucleus may also contribute to the inhibitory actions of 15d-PGJ$_2$ on the NF-κB pathway at later treatment times. This possibility is also supported by the observation of an attenuation in the cytokine-elicited decrease in the cytosolic levels of IκBα in the cells that had been pre-treated with 15d-PGJ$_2$. After 15 min of cytokine stimulation, levels of IκBα were 1.7-fold higher in cells activated in the presence of 15d-PGJ$_2$ than in cells treated with cytokines alone (Figure 5C). These results suggest that both, an impairment of nuclear translocation of NF-κB subunits and a direct inhibition of NF-κB-DNA complex formation may contribute to the reduction in the NF-κB DNA binding activity elicited by 15d-PGJ$_2$ in RMC.

15d-PGJ$_2$ Inhibits NF-κB DNA Binding Activity in Vitro, without Disrupting NF-κB Dimers

To evidence a potential direct effect of cyPG on NF-κB-DNA complex formation we analyzed the effect of 15d-PGJ$_2$ on NF-κB DNA binding activity *in vitro* by EMSA. Confluent RMC were activated with Ck (IL-1β/TNFα) in serum-free medium during 15 minutes. Four micrograms of the nuclear extracts isolated as described in Methods, were incubated with different concentrations of prostaglandin before the binding assay. As observed in Figure 6A, treatment of nuclear extracts with 15d-PGJ$_2$ *in vitro* effectively inhibited NF-κB-DNA complex formation (43% inhibition in nuclear extracts treated with 10 μM 15d-PGJ$_2$ *versus* control).

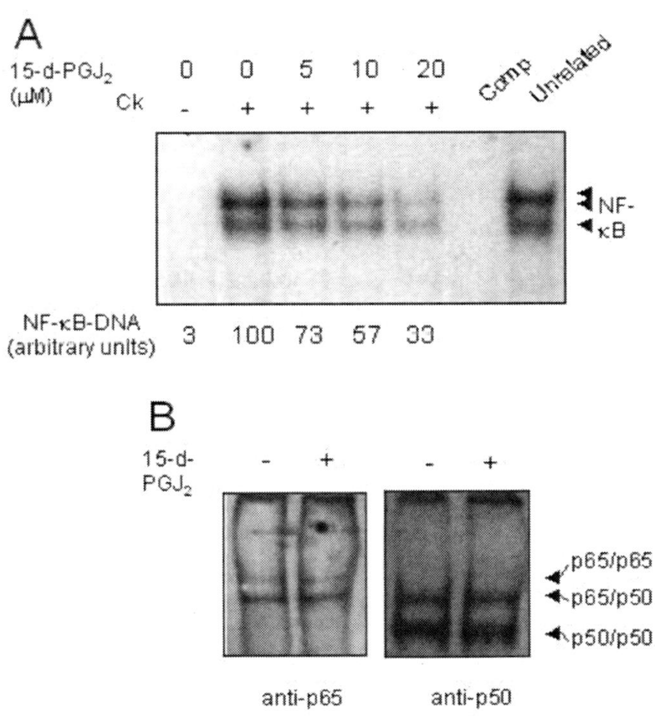

Figure 6. Effect of 15d-PGJ$_2$ on NF-κB DNA binding *in vitro*. A) RMC were activated with Ck for 15 min. Nuclear extracts were isolated and incubated with the indicated concentrations of 15d-PGJ$_2$ for 1 hour at r.t. The NF-κB-DNA binding activity was analyzed by EMSA. Competition assays were carried out using an excess of unlabeled NF-κB or unrelated oligonucleotide (200-fold). Protein-DNA complexes were quantitated by image scanning and the mean values from three experiments are shown. B) RMC were activated as before and the nuclear extracts were incubated in presence or absence of 5 μM 15d-PGJ$_2$ as above. The integrity of the dimers was analyzed by electrophoresis under non-denaturing conditions and Western blot with specific antibodies against p65 and p50. Results shown are representative of three experiments.

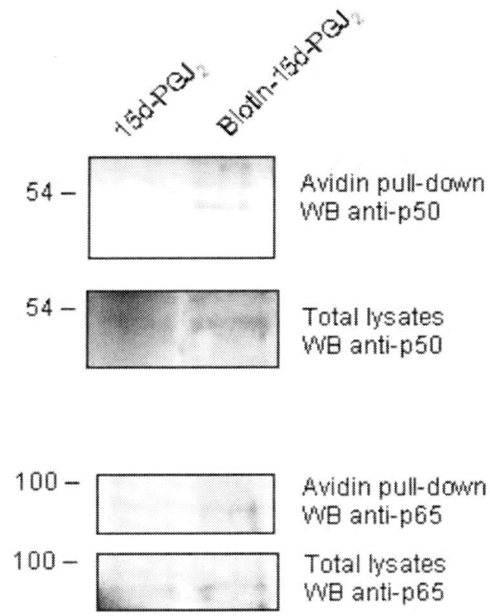

Figure 7. Biotinylated 15d-PGJ_2 binds to NF-κB proteins in RMC. RMC were incubated with 10 μM 15d-PGJ_2 or biotinylated 15d-PGJ_2 in serum-free medium for 2 h. Cell lysates were incubated with avidin beads. After extensive washing, proteins retained on the beads were eluted in Laemmli buffer and analyzed by gel electrophoresis and Western blot with antibodies against the p50 and p65 subunits of NF-κB, as indicated. The levels of NF-κB subunits in total cell lysates are shown. Results are representative of two experiments.

To discard the possibility that the inhibitory effect could be due to a disruption or degradation of the NF-κB dimers present in the nuclear extracts during the incubation with the PG, we performed an immunoblot of the non-denaturing gels run under the same conditions used for EMSA with specific antibodies against p50 and p65 subunits (Figure 6B). This Western blot showed the presence of protein bands which were detected only by anti-p65 antibodies, corresponding probably to p65 homodimers. A faster migrating band reacting with both anti-p65 and anti-p50 antibodies was also detected which was probably formed by p65/p50 heterodimers. Finally, the lower band, reacting only with anti-p50 antibodies was probably due to p50 homodimers. The p65/p50 and the p50 bands co-migrated with the main specific bandshifts observed in EMSAs performed with the labeled oligonucleotide. As it can be observed in Figure 6B, incubation of the isolated nuclear extracts in the presence of 15d-PGJ_2, in the conditions under which NF-κB-DNA binding was clearly inhibited, did not affect the integrity of the NF-κB dimers. These results indicate that the effect of 15d-PGJ_2 is probably due to an impairment in the formation of NF-κB-DNA complexes.

Biotinylated 15d-PGJ_2 Binds to p50 and p65 Proteins in Intact RMC

We have previously shown that 15d-PGJ_2 may interact with p50 *in vitro* [62]. To explore whether 15d-PGJ_2 may directly interact with NF-κB proteins in RMC we incubated intact cells in the presence of a biotinylated 15d-PGJ_2 derivative (Figure 7). This compound retains the cyclopentenone moiety of the PG and it is able to modify cysteine residues in proteins by Michael addition (see Figure 3).

After incubation with biotinylated 15d-PGJ_2, proteins with biotin moieties were collected by incubation with avidin-agarose beads and the proteins of interest were detected in the eluate by Western blot. This type of assay showed the presence of both p50 and p65 NF-κB subunits in the avidin-binding fraction of RMC incubated with biotinylated 15d-PGJ_2 but not in the control samples from RMC incubated with non-biotinylated PG. These results suggest that biotinylated 15d-PGJ_2 is able to form covalent adducts with NF-κB proteins in RMC.

Incorporation of Biotinylated cyPG into the NF-κB Subunit p50 in Vitro Is Time and Concentration-Dependent

We were then interested in exploring the structural determinants involved in the interaction between NF-κB proteins and cyPG. For this purpose we have set up the conditions for a Western blot-based assay that would allow us to monitor the *in vitro* modification of p50 by biotinylated cyPG in a semi-quantitative manner. As it is shown in Figure 8, the incorporation of biotinylated 15d-PGJ_2 into the p50 construct was linear with time at least up to 4 h incubation.

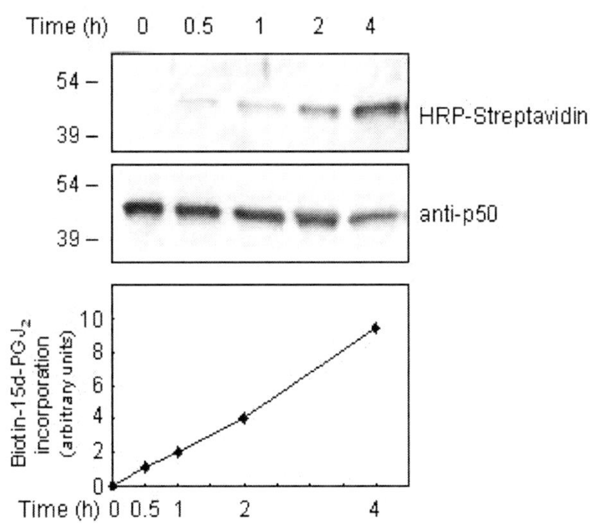

Figure 8. Time-course of the incorporation of biotinylated 15d-PGJ_2 into p50. Human recombinant p50 construct at 0.5 μM was incubated with 1 μM biotinylated 15d-PGJ_2 at r.t. for the indicated time periods. Incubation mixtures were analyzed by SDS-PAGE and Western blot and incorporation of biotin was detected with HRP-conjugated streptavidin and ECL. The amount of p50 present in the incubation was assessed by Western blot with anti-p50 antibody. The intensity of the biotin signal, normalized by the signal given by the anti-p50 antibody, and expressed in arbitrary units, is shown in the lower panel. Results are representative of three assays.

Figure 9 shows the concentration-dependence of this assay. Incorporation of biotinylated cyPG was dependent on the concentration of the PG (Figure 9A). Interestingly, under the same conditions, the modification of p50 by biotinylated 15d-PGJ_2 was more intense than with biotinylated PGA_1. This difference may be due to the different reactivity of the two PG or to factors related to the structural features of the protein. As it is shown in Figure 9B, 15d-PGJ_2 binding to p50 in this assay was also linear with respect to the concentration of the p50 construct. This indicates that this type of assay provides a semi-quantitative approach to study the factors which influence the binding of biotinylated cyPG to p50 *in vitro*.

Binding of Biotinylated 15d-PGJ_2 to p50 is Selective

In a previous study we have reported that *in vitro* treatment of p50 with 15d-PGJ_2 strongly diminishes its DNA binding ability [62]. PGA_1 exerted a moderate inhibitory effect, while carbaprostacyclin and PGE_2, which are not cyclopentenones, did not inhibit the binding. Here we have used the Western blot assay to explore the ability of several PG to block the binding of biotinylated 15d-PGJ_2 to p50. The results depicted in Figure 10 show that the binding of biotinylated 15d-PGJ_2 to p50 can be blocked by an excess of 15d-PGJ_2. Excess PGA_1 partially reduced labeling while PGE had no effect. These observations suggest that the binding of 15d-PGJ_2 to p50 is selective and that the cyclopentenone moiety is an important determinant for the interaction.

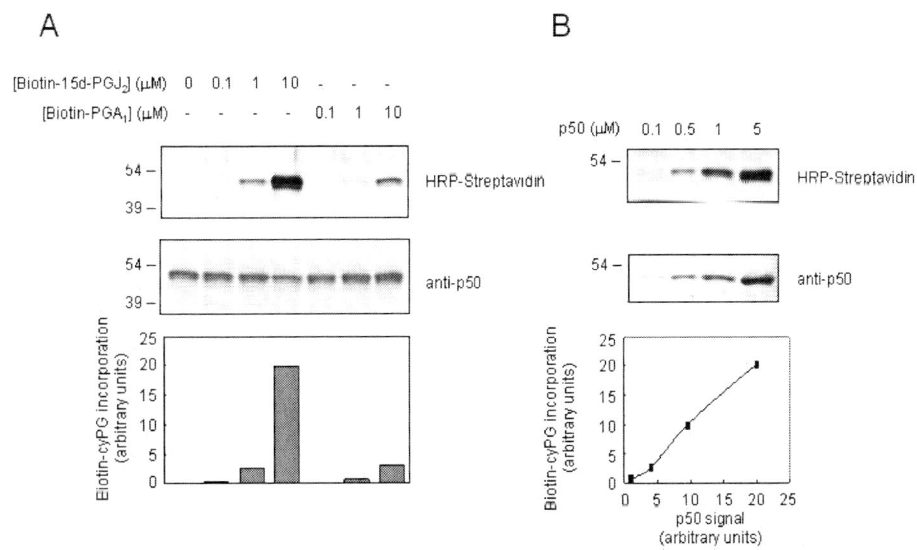

Figure 9. Dose-response of the incorporation of biotinylated cyPG into p50. (A) Recombinant p50 at 0.5 μM was incubated in the presence of the indicated concentrations of biotinylated cyPG for 1 h at r.t. Biotinylated cyPG incorporation was estimated as described in the legend of Figure 8. (B) The indicated concentrations of p50 were incubated with 1 μM biotinylated 15d-PGJ_2 at r.t. for 1 h. The lower panel shows the plot of the biotin signal versus the p50 signal. Results are representative of two assays.

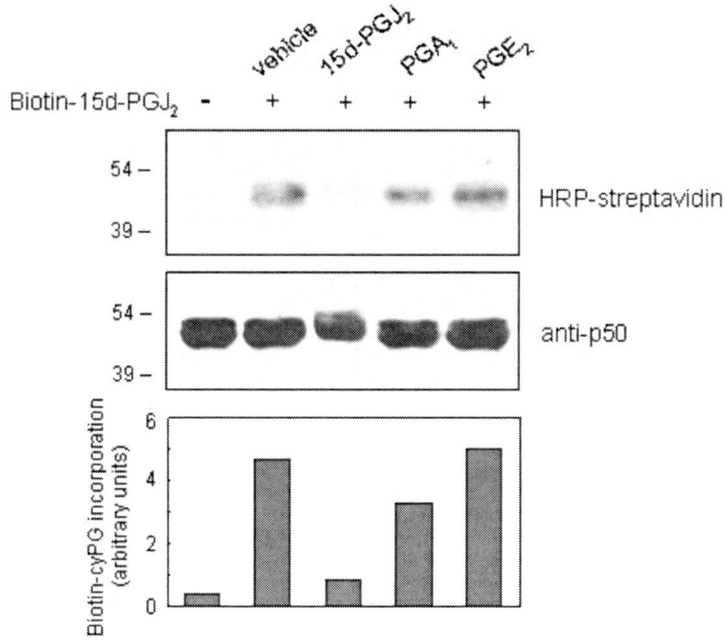

Figure 10. Effect of several PG on the binding of biotinylated 15d-PGJ$_2$ to p50. p50 at 0.5 μM was incubated with 1 μM biotinylated 15d-PGJ$_2$ in the absence or presence of the indicated PG at 100 μM. Biotin incorporation was assessed as above. Results are representative of two assays.

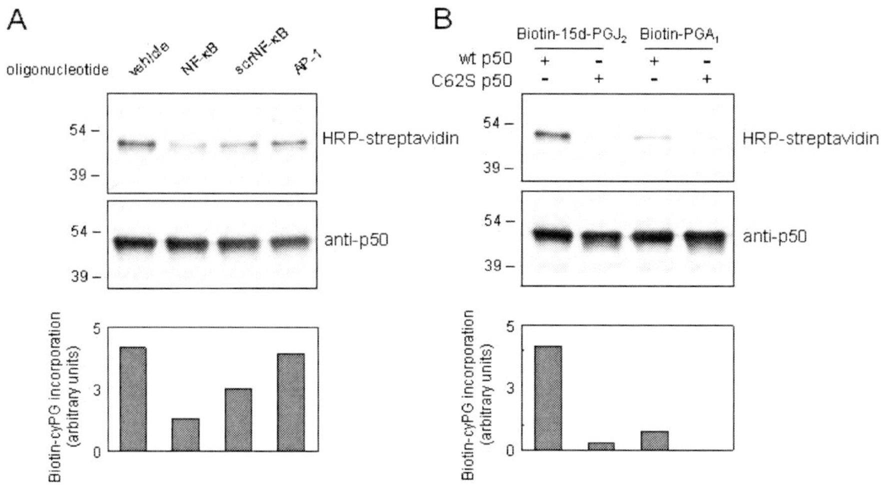

Figure 11. Involvement of the DNA binding region of p50 in the interaction with biotinylated cyPG. (A) Effect of different oligonucleotides on the binding of biotinylated 15d-PGJ$_2$ to p50. p50 at 0.5 μM was incubated in the presence of the indicated oligonucleotides or annealing buffer (vehicle) for 15 min at r.t., after which, biotinylated 15d-PGJ$_2$ at 1 μM final concentration was added and the incubation was continued for 1 h. NF-κB, 5'-GGAGAGGGGATTCCCTGCG-3'; scrNF-κB, 5'- GGACGAGCTGCGT GCGTGA-3'; AP-1, 5'-TGTTGGTGACTAATAACAC-3'. (B) Wild type (wt) or cysteine 62 to serine mutant p50 (C62S) at 0.5 μM was incubated with 1 μM biotinylated 15d-PGJ$_2$ or PGA$_1$. Biotin incorporation and p50 levels were assessed as described in the legend of Figure 8.

Biotinylated cyPG Bind to Cysteine 62 in Human p50, in a Manner Dependent on the Interaction with DNA

The ability of cyPG to inhibit NF-κB-DNA binding suggests that binding of cyPG directly precludes the interaction of p50 with DNA. We then explored whether p50 binding to DNA could affect the incorporation of biotinylated 15d-PGJ_2. As it is shown in Figure 11A, incubation of p50 with a double stranded oligonucleotide bearing the sequence of one of the NF-κB binding sites from the human COX-2 promoter, as detailed in Methods, inhibited the binding of biotinylated 15d-PGJ_2 by approximately 70%. In contrast, an NF-κB oligonucleotide with a mutated sequence elicited only a 40% inhibition, while a nucleotide with a consensus sequence for the binding of AP-1 was virtually without effect. Therefore, the interaction of p50 with DNA exerts a protective effect on the modification of critical cysteine residues by cyPG. This suggests that the interaction of 15d-PGJ_2 with p50 involves residues closely implicated in DNA binding. Evidence obtained in previous work strongly suggested that cysteine 62 of human p50, which is located in the DNA binding domain of this protein, was involved in the interaction with 15d-PGJ_2 [62, 67]. Here we have explored whether this cysteine residue plays also a role in the interaction of p50 with biotinylated PGA_1. As it is shown in Figure 11B, the incorporation of both biotinylated 15d-PGJ_2 and PGA_1 into p50 was strongly reduced (more than 90% inhibition) when cysteine 62 was mutated to serine. These results indicate that cysteine 62 is an important target for the binding of both A-type and J-type cyPG to p50.

CyPG Interact with AP-1 Proteins

Evidence available indicate that the cellular targets of cyPG are multiple. We were then interested in exploring the presence of other potential targets for modification by cyPG in RMC. Incubation of intact RMC with biotinylated 15d-PGJ_2 resulted in the incorporation of the biotin label into multiple bands, as detected by gel electrophoresis and Western blot with HRP-conjugated streptavidin (Figure 12A). The bands detected were clearly different from the endogenous biotinylated proteins. This indicates that the modification of several potential protein targets may contribute to the effects of cyPG in RMC. Among these targets we have recently reported that the component of AP-1 c-Jun can be modified by biotinylated 15d-PGJ_2 in murine fibroblasts [59]. 15d-PGJ_2 has been shown to inhibit the activity of AP-1 in several experimental systems [48, 68], and this effect could contribute to the antiinflammatory actions of cyPG. The potential modification of AP-1 by biotin-15d-PGJ_2 in mesangial cells was explored by means of an avidin pull-down assay. As it is shown in Figure 12B, c-Jun present in RMC lysates was retained on avidin beads when cells had been incubated with biotinylated 15d-PGJ_2 but not when cells had been treated with non biotinylated PG. This indicates that in this cell type, AP-1 components may also be targets for cyPG.

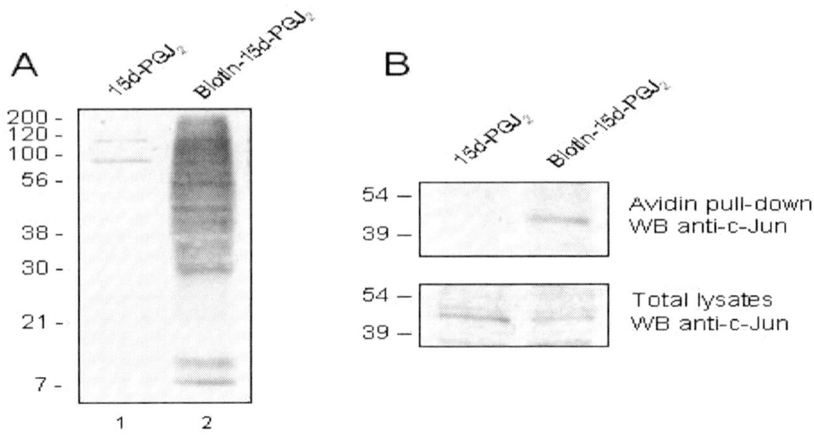

Figure 12. The transcription factor AP-1 is a target for biotinylated 15d-PGJ$_2$ binding in RMC. (A) RMC were incubated for 2 h with 10 μM 15d-PGJ$_2$ or biotinylated 15d-PGJ$_2$. Aliquots from cell lysates containing 10 μg of protein were analyzed by SDS-PAGE, Western blot and detection with HRP-streptavidin. Biotin positive polypeptides were visualized by ECL. Bands detected in lane 1 correspond to endogenous biotinylated proteins. (B) Cell lysates were incubated with avidin beads. After extensive washing, proteins retained on the beads were eluted with Laemmli sample buffer. The presence of c-Jun in the eluates was assessed by Western blot and detection with HRP-conjugated streptavidin and ECL. Results shown are representative of three assays.

Discussion

Evidence accumulated during the last several years indicate that cyPG at micromolar concentrations exert potent antiinflammatory effects both in cellular systems and in animal models of inflammation. The mechanisms responsible for these effects appear to be multiple and they have been only partially characterized. Inhibition of signal transduction pathways by cyPG and direct interference with transcription factors have been reported. Given their importance for the inflammatory response, the transcription factors NF-κB and AP-1 have been studied as potential targets for the antiinflammatory actions of cyPG. Inhibitory effects of cyPG on NF-κB activity have been previously described, and they have been shown to correlate with the antiinflammatory potency of cyPG *in vivo*. Inhibition of the expression of proinflammatory genes in association with NF-κB blockade has been proposed to underlie the beneficial effects of cyPG treatment on acute renal failure, restenosis after balloon angioplasty in rats or carrageenin-induced pleurisy [69-71]. In this chapter we have described that the cyPG 15d-PGJ$_2$ reduces the transcriptional activity of NF-κB elicited by cytokines in RMC, as assessed by transfection of a reporter plasmid. CyPG and in particular 15d-PGJ$_2$ may interfere with the NF-κB activity at multiple levels. The two main mechanisms characterized to date are related to the ability of cyPG to activate PPAR or to modify cellular proteins by Michael addition. Several proteins along the NF-κB activation pathway have been proposed to be targets for modification by cyPG. CyPG has been reported to directly impair the function of activated IKK both in intact cells and *in vitro*, reducing its ability to phosphorylate IκB which results in decreased degradation of IκB and impaired nuclear translocation of NF-κB dimers [72, 73]. The requirement of a cysteine residue at position 179 in the activation loop of IKKβ for this inhibition to occur, suggested the interaction between the cyclopentenone moiety of the cyPG and the sulphydryl group of this residue by Michael addition. More recently, adducts of biotinylated cyPG and both, IKKα and β have been reported [74, 75]. The contribution of this interaction to the inhibition of NF-κB appears to depend on the cell system under study. In HeLa cells, inhibition of NF-κB transcriptional activity and DNA binding does not involve the inhibition of IKK activity, IκB degradation or nuclear translocation of NF-κB subunits [61, 62]. In contrast, in murine macrophages cyPG impair IκBα phosphorylation and degradation, thus pointing to the involvement of the inhibition of IKK [61, 72]. In addition cyPG can interfere with the activity of the proteasome [76], and a biotinylated 15d-PGJ_2-proteasome conjugate has been recently identified [77]. The results presented in this chapter suggest that in RMC cyPG may impair IκBα degradation, since the reduction in IκBα levels elicited by cytokines is attenuated in cells activated in the presence of 15d-PGJ_2. However, whether this effect is due to an inhibition of IκB phosphorylation by IKK or to decreased proteasome activity requires further investigation. In addition, our results show that 15d-PGJ_2 can inhibit the DNA binding activity of NF-κB more intensely than the nuclear translocation of NF-κB subunits p65 and p50, and that this inhibition can also occur in isolated nuclear extracts. These results suggest that 15d-PGJ_2 can directly interfere with NF-κB DNA binding in RMC. These results are in agreement with previous reports in which cyclopentenones have been shown to block the binding of recombinant p65 [61] or p50 [62] to DNA. The requirement for the presence of specific cysteine residues, cysteine 38 in human p65 and cysteine 62 in human p50, for this inhibition to occur, points to the involvement of a direct interaction between the cyclopentenone moiety of the PG and the sulphydryl groups in proteins. In this chapter we provide evidence for the formation of covalent adducts between biotinylated 15d-PGJ_2 and p65 and p50 in intact RMC. These results confirm previous observations from our group in HeLa cells [62]. A schematic view of the NF-κB activation pathway with the potential sites of interaction of cyPG is shown in Figure 13. In addition, we have monitored this interaction by a semi-quantitative assay using recombinant p50 *in vitro*. Under the conditions of our assay the modification of p50 by 15d-PGJ_2 can be detected with concentrations of the PG as low as 100 nM, and it is more efficient than the modification by PGA_1. At present it is not known whether this differential effect might be due to structural determinants of the PG or the protein. This will be the subject of future investigations. The modification of p50 by cyPG *in vitro* appears to involve a region which is masked by DNA binding, since incubation in the presence of an oligonucleotide with the consensus sequence for NF-κB binding reduces incorporation of the PG. The cysteine residue located at position 62 in the sequence of human p50 has been shown to be important for DNA binding. Modification of this residue by oxidation or alkylating agents results in loss of DNA binding activity [34]. Substitution of this cysteine residue by serine renders the p50 protein virtually insensitive to redox-induced inhibition of DNA binding [34]. Interestingly, under our experimental conditions, the incorporation of biotinylated cyPG into the cys62ser mutant is reduced by more than 90%. This indicates that this cysteine residue is the main site of attachment for both 15d-PGJ_2 and PGA_1 derivatives *in vitro*.

We and other authors have reported that cyPG can interact with the AP-1 pathway by PPAR-dependent and independent mechanisms [59, 60]. These interactions may result either in activation of inhibition of AP-1 depending on the cell model under study. Treatment of RMC with 15d-PGJ$_2$ has been reported to result in a reduction in AP-1 DNA binding [59, 68]. Our results show that in RMC the component of the AP-1 transcription factor c-Jun may also be a target for modification by biotinylated 15d-PGJ$_2$. This suggests that inhibition of AP-1 by direct modification with cyPG could contribute to the antiinflammatory effects of cyPG in RMC.

Not only NF-κB/Rel proteins and c-Jun, but other transcription factors involved in the inflammatory response possess critical cysteine residues susceptible to redox regulation that could be targeted by direct modification with cyPG. Examples of these factors are cAMP response element binding protein (CREB), glucocorticoid receptor and hypoxia inducible factor 1α (HIF-1α), among others. Therefore, the possibility exists that cyPG can form covalent adducts with other transcription factors not yet identified. In addition, as detailed above for NF-κB, the function of transcription factors important for the cellular defense against stress or for inflammation can be regulated by the direct modification not of the transcription factors themselves but of cytosolic or nuclear proteins involved in the regulation of their activity. This is the case of the protein Keap1, which controls the subcellular localization and degradation of the transcription factor Nrf2, involved in the regulation of genes encoding antioxidant enzymes. Modification of Keap1 by cyPG leads to activation of Nrf2 [78, 79]. This type of indirect mechanism could also influence the activity of the factors subject of this study, since both thioredoxin and thioredoxin reductase, proteins involved in the maintenance of the reduced state of critical cysteine residues in AP-1, Rel and p53 proteins, have been shown to be modified and inhibited by cyPG [75, 80].

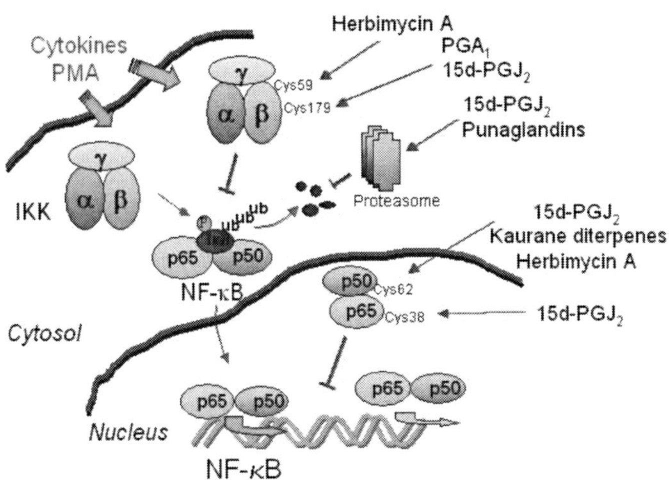

Figure 13. Schematic view of the NF-κB activation pathway showing the sites of interaction of several Michael receptors. Given the importance of the transcription factor NF-κB for the inflammatory response, the pathway leading to its activation has been the subject of numerous studies in search for antiinflammatory molecules. Shown is the site of action of several molecules which have been reported to interfere with NF-κB activation through the formation of covalent adducts with critical proteins of this pathway by Michael addition.

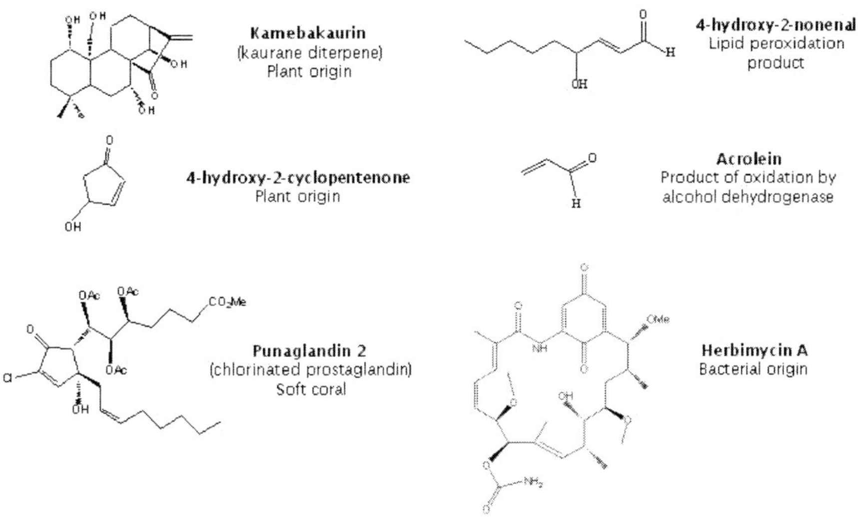

Figure 14. Structure of Michael receptors of diverse origin

In the light of the importance of covalent protein modification in the antiinflammatory effects of cyPG, it is interesting to consider the potential of other compounds with unsaturated carbonyl groups that could act by similar mechanisms. This is the case of several compounds of plant origin, the structure of some of which is shown in Figure 14. The kaurane diterpene kamebakaurin is present in plants used in oriental medicine for the treatment of cancer and inflammation. The mechanism of action of this compound has been reported to involve the formation a covalent adduct with the p50 subunit of NF-κB, targeting cysteine 62 of p50 in a manner analogous to 15d-PGJ_2 [81]. The compound 4-hydroxy-2-cyclopentenone is present in the extracts of *Passiflora tetranda* [82], and it is responsible for the antibacterial properties of this preparation which has been used in herbal medicine. Recently, a class of compounds termed punaglandins has been isolated from the soft coral *Telesto riisei* [83], and they have been reported to exhibit potent antiinflammatory and antiproliferative effects. The molecular basis for these actions may rely, at least in part, in their ability to function as potent Michael receptors to inhibit the ubiquitin isopeptidase activity of the proteasome. In addition, several products of lipid peroxidation such as 4-hydroxy-2-nonenal or acrolein possess unsaturated carbonyl groups that may form covalent adducts with proteins. Proteins carrying these oxidative modifications may be more resistant to degradation and accumulate in cells, thus contributing to the pathophysiology of various disease processes including atherosclerosis, diabetes or Alzheimer's disease [84, 85]. Unsaturated carbonyl groups are also present in certain drugs and enzyme inhibitors, like ethacrynic acid or herbimycin. The tyrosine kinase inhibitor herbimycin A has been found to block NF-κB activation by IL-1 or phorbol 12-myristate 13-acetate (PMA) in several cell types through the interaction with cysteine 62 in p50 [86]. Recently, the interaction of herbimycin with the β subunit of IKK has been reported [87]. This interaction, which abrogates IKK activity and NF-κB activation has been reported to occur at cysteine 59, which is located near the catalytic site of IKK. Therefore, in a manner analogous to cyPG, herbimycin A can also target several steps in the NF-κB activation pathway and for this reason it has been proposed as a novel candidate for antiinflammatory drug. The sites of action of some Michael receptors along the NF-κB activation pathway is shown in Figure 13. Interestingly, inhibition of IKKβ by various agents has been reported to result in lower GSH content and redox potential [88]. This could constitute an additional mechanism of action for cyPG and other molecules targeting IKKβ that could lead to the modulation of imflammation through alteration of the cellular redox status. In recent years, both proteomic and hypothesis-driven approaches have greatly widened our perspective of the targets and effects of cyPG [89, 90]. In these studies it has been shown that cyPG may target a broad but selective set of cellular proteins involved in structural and regulatory functions [91-93]. Moreover, the modification of proteins by cyPG shows both intra-molecular and inter-molecular selectivity, which depends of structural features of the protein and the cyPG, along with context factors [94]. Interestingly, it is becoming clear that a reciprocal regulation exists between protein modification by cyPG and cellular redox status [95, 96].

In summary, cyPG represent a class of compounds with high reactivity towards cellular thiols which exert varied biological actions of potential pharmacological interest. Among the targets of cyPG, the NF-κB and AP-1 activation pathways constitute important points of intervention for the control of inflammatory processes. Elucidation of the mechanisms underlying the effects of cyPG and identification of other potential targets will contribute to assess their potential as antiinflammatory or antiviral molecules.

Acknowledgments

We wish to acknowledge the work of collaborators who have contributed to the research on this topic, Francisco J. Sánchez-Gómez, Konstatinos Stamatakis, Dr. E. Pineda-Molina, Dr. J. Gayarre, B.Garzón, B. Díez-Dacal and Dr. F. J. Cañada, and the technical assistance of M. J. Carrasco. Our work is supported by Grant SAF2009-11642 from Ministerio de Ciencia e Innovación and RETIC RIRAAF 07/0064/0007 from Instituto de Salud Carlos III.

Abbreviations

PG	prostaglandin(s);
CyPG	cyclopentenone prostaglandin(s);
COX	cyclooxygenase;
15d-PGJ_2	15-deoxy-$\Delta^{12,14}$-PGJ_2;
NF-κB	nuclear factor-kappaB;
IKK	IkappaB kinase;
PPAR	peroxisome proliferator activated receptor;
AP-1	activator protein 1;
STAT-1	signal transducer and activator of transcription 1;
CREB	cAMP response element binding protein;
iNOS	inducible nitric oxide synthase;
NO	nitric oxide;
GSH	reduced glutathione;
HIF	hypoxia inducible factor;
RMC	rat mesangial cells;

HRP	horseradish peroxidase;
Ck	cytokines;
IL-1β	interleukin 1-beta;
TNF-α	tumor necrosis factor alpha;
LPS	bacterial lipopolysaccharide;
EMSA	electrophoretic mobility shift assay;
PMA	phorbol 12-myristate 13-acetate;
ECL	enhanced chemiluminiscence.

REFERENCES

[1] Nathan, C. (2002). Points control in inflammation. *Nature*, *420*, 846-852.

[2] Kunz, D., Walker, G., Eberhardt, W. & Pfeilschifter, J. (1996). Molecular mechanisms of dexamethasone inhibition of nitric oxide synthase expression in interleukin 1β-stimulated mesangial cells: evidence for the involvement of transcriptional and posttranscriptional regulation. *Proc. Natl. Acad. Sci.* USA, *93*, 255-259.

[3] Laskin, J. D., Heck, D. E. & Laskin, D. L. (1994). Multifunctional role of nitric oxide in inflammation. *Trends Endocrinol. Metab.*, *5*, 377-382.

[4] Pérez-Sala, D. & Lamas, S. (2001). Regulation of cyclooxygenase-2 expression by nitric oxide in cells. *Antioxid. Redox Signal.*, *3*, 231-248.

[5] Saura, M., Martínez-Dalmau, R., Minty, A., Pérez-Sala, D. & Lamas, S. (1996). Interleukin-13 inhibits inducible nitric oxide synthase expression in human mesangial cells. *Biochem. J.*, *313*, 641-646.

[6] Pérez-Sala, D., Díaz-Cazorla, M., Ros, J., Jiménez, W. & Lamas, S. (1997). Tetrahydrobiopterin modulates cyclooxygenase-2 expression in human mesangial cells. *Biochem. Biophys. Res. Commun.*, *241*, 7-12.

[7] Saura, M., Zaragoza, C., Díaz-Cazorla, M., Hernández-Perera, O., Eng, E., Lowenstein, C. J., Pérez-Sala, D. & Lamas, S. (1998). Involvement of transcriptional mechanisms in the inhibition of NOS2 expression by dexametnasone in rat mesangial cells. *Kidney Int.*, *53*, 38-49.

[8] Tetsuka, T., Daphna-Iken, D., Srivastava, S. K., Baier, L. D., DuMaine, J. & Morrison, A. R. (1994). Cross-talk between cyclooxygenase and nitric oxide pathways: prostaglandin E_2 negatively modulates induction of nitric oxide synthase by interleukin 1. *Proc. Natl. Acad. Sci. USA*, *91*, 12168-12172.

[9] Mühl, H. & Pfeilschifter, J. (1994). Tetrahydrobiopterin is a limiting factor of nitric oxide generation in interleukin 1ß-stimulated rat glomerular mesangial cells. *Kidney Int.*, *46*, 1302-1306.

[10] Tetsuka, T., Daphna-Iken, D., Miller, B. W., Guan, Z., Baier, L. D. & Morrison, A. R. (1996). Nitric oxide amplifies interleukin 1-induced cyclooxygenase-2 expression in rat mesangial cells. *J. Clin. Invest.*, *97*, 2051-2056.

[11] Díaz-Cazorla, M., Pérez-Sala, D. & Lamas, S. (1999). Dual effect of nitric oxide donors on cyclooxygenase-2 expression in human mesangial cells. *J. Am. Soc. Nephrol.*, *10*, 943-952.

[12] Schlöndorff, D. (1996). Roles of the mesangium in glomerular function. *Kidney Int.*, *49*,

1583-1585.

[13] Mené, P. (1996). Physiology and pathophysiology of the mesangial cell. *Nefrología, 16*, 8-13.

[14] Rzymkiewicz, D., Leingang, K., Baird, N. & Morrison, A. (1994). Regulation of prostaglandin endoperoxide synthase gene expression in rat mesangial cells by interleukin-1 beta. *Am. J. Physiol., 266*, F39-F45.

[15] Saura, M., López, S., Rodríguez-Puyol, M., Rodríguez-Puyol, D. & Lamas, S. (1995). Regulation of inducible nitric oxide synthase expression in rat mesangial cells and isolated glomeruli. *Kidney Int., 47*, 500-509.

[16] Salvemini, D., Settle, S. L., Masferrer, J. L., Seibert, K., Currie, M. G. & Needleman, P. (1995). Regulation of prostaglandin production by nitric oxide; an in vivo analysis. *Br. J. Pharmacol., 114*, 1171-1178.

[17] Salvemini, D. & Masferrer, J. L. (1996). Interactions of nitric oxide with cyclooxygenase: in vitro, ex vivo and in vivo studies. *Methods Enzymol., 269,* 12-25.

[18] Nogawa, S., Forster, C., Zhang, F., Nagayama, M., Ross, M. E. & Iadecola, C. (1998). Interaction between inducible nitric oxide synthase and cyclooxygenase- 2 after cerebral ischemia. *Proc. Natl. Acad. Sci. U S A, 95*, 10966-10971.

[19] von Knethen, A. & Brüne, B. (1997). Cyclooxygenase-2: an essential regulator of NO-mediated apoptosis. *FASEB J., 11*, 887-895.

[20] von Knethen, A. & Brune, B. (2000). Superinduction of cyclooxygenase-2 by NO(*) and agonist challenge involves transcriptional regulation mediated by AP-1 activation. *Biochemistry, 39*, 1532-1540.

[21] Habib, A., Bernard, C., Lebret, M., Créminon, C., Esposito, B., Tedgui, A. & Maclouf, J. (1997). Regulation of the expression of cyclooxygenase-2 by nitric oxide in rat peritoneal macrophages. *J. Immunol., 158*, 3845-3851.

[22] Pérez-Sala, D., Cernuda-Morollón, E., Díaz-Cazorla, M., Rodríguez-Pascual, F. & Lamas, S. (2001). Posttranscriptional regulation of human iNOS by the NO/cGMP pathway. *Am. J. Physiol. Renal Physiol., 280*, F466-F473.

[23] Marotta, P., Sautebin, L. & Di Rosa, M. (1992). Modulation of the induction of nitric oxide synthase by eicosanoids in the murine macrophage cell line J774. *Br. J. Pharmacol., 107*, 640-641.

[24] Gilroy, D. W., Colville-Nash, P. R., Willis, D., Chivers, J., Paul-Clark, M. J. & Willoughby, D. A. (1999). Inducible cyclooxygenase may have anti-inflammatory properties. *Nat. Med., 5*, 698-701.

[25] Levy, B. D. (2001). Lipid mediator class switching during acute inflammation: signals in resolution. *Nat. Immunol., 2*, 612-619.

[26] Karin, M. (1995). The regulation of AP-1 activity by mitogen-activated protein kinases. *J. Biol. Chem., 270*, 16483-16486.

[27] Ghosh, S., May, M. J. & Kopp, E. B. (1998). NF-κB and rel proteins: evolutionarily conserved mediators of immune responses. *Annu. Rev. Immunol., 16*, 225-260.

[28] Baeuerle, P. A. (1998). Pro-inflammatory signaling: Last pieces in the NF-κB puzzle? *Curr. Biol., 8*, R19-R22.

[29] Ajuebor, M. N., Virag, L., Flower, R. J., Perretti, M. & Szabo, C. (1998). Role of inducible nitric oxide synthase in the regulation of neutrophil migration in zymosan-induced inflammation. *Immunology, 95*, 625-630.

[30] Connelly, L., Palacios-Callender, M., Ameixa, C., Moncada, S. & Hobbs, A. J. (2001). Biphasic regulation of NF-kappaB activity underlies the pro- and anti-inflammatory actions of nitric oxide. *J. Immunol.*, *166*, 3873-3881.

[31] Peng, H.-B., Libby, P. & Liao, J. K. (1995). Induction and stabilization of IκBα by nitric oxide mediates inhibition of NF-κB. *J. Biol. Chem.*, *270*, 14214-14219.

[32] Katsuyama, K., Shichiri, M., Marumo, F. & Hirata, Y. (1998). NO inhibits cytokine-induced iNOS expression and NF-kappaB activation by interfering with phosphorylation and degradation of IkappaB-alpha. *Arterioscler. Thromb. Vasc. Biol.*, *18*, 1796-11802.

[33] Mathews, J. R., Botting, C. H., Panico, M., Morris, H. R. & Hay, R. T. (1996). Inhibition of NF-kappaB DNA binding by nitric oxide. *Nucl. Acid Res.*, *24*, 2236-2242.

[34] Pineda-Molina, E., Klatt, P., Vázquez, J., Marina, A., Lacoba, M. G. d., Pérez-Sala, D. & Lamas, S. (2001). Glutathionylation of the p50 subunit of NF-κB: a mechanism for redox-induced inhibition of DNA-binding. *Biochemistry*, *40*, 14134-14142.

[35] Shirakawa, F. & Mizel, S. (1989). In vitro activation and nuclear translocation of NF-kappaB catalyzed by cyclic AMP-dependent protein kinase and protein kinase C. *Mol. Cell. Biol.*, *9*, 2424-2430.

[36] Zhong, H., Yang, H. S., Erdjument-Bromage, H., Tempst, P. & Ghosh, S. (1997). The transcriptional activity of NF-kappaB is regulated by the IkappaB- associated PKAc subunit through a cyclic AMP-independent mechanism. *Cell*, *89*, 413-424.

[37] Poligone, B. & Baldwin, A. S. (2001). Positive and negative regulation of NF-kappaB by COX-2: roles of different prostaglandins. *J. Biol. Chem.*, *276*, 38658-38664.

[38] Nusing, R. M., Klein, T., Pfeilschifter, J. & Ullrich, V. (1996). Effect of cyclic AMP and prostaglandin E_2 on the induction of nitric oxide- and prostanoid-forming pathways in cultured rat mesangial cells. *Biochem. J.*, *313*, 617-623.

[39] D'Acquisto, F., Sautebin, L., Iuvone, T., Di Rosa, M. & Carnuccio, R. (1998). Prostaglandins prevent inducible nitric oxide synthase protein expression by inhibiting nuclear factor-kappaB activation in J774 macrophages. *FEBS Lett.*, *440*, 76 80.

[40] Shibata, T., Kondo, M., Osawa, T., Shibata, N., Kobayashi, M. & Uchida, K. (2002). 15-Deoxy-Delta 12,14-prostaglandin J2. A prostaglandin D2 metabolite generated during inflammatory processes. *J. Biol. Chem.*, *277*, 10459-10466.

[41] Kondo, M., Shibata, T., Kumagai, T., Osawa, T., Shibata, N., Kobayashi, M., Sasaki, S., Iwata, M., Noguchi, N. & Uchida, K. (2002). 15-Deoxy-Delta 12,14-prostaglandin J2: The endogenous electrophile that induces neuronal apoptosis. *Proc. Natl. Acad. Sci. U S A*, *99*, 7367-7372.

[42] Thomas, S. C., Ryan, M. A., Shanley, T. P. & Wong, H. R. (1998). Induction of the stress response with prostaglandin A_1 increases IκBα gene expression. *FASEB J.*, *12*, 1371-1378.

[43] Santoro, M. G. (2000). Heat shock factors and the control of the stress response. *Biochem. Pharmacol.*, *59*, 55-63.

[44] Forman, B. M., Tontonoz, P., Chen, J., Brun, R. P., Spiegelman, B. M. & Evans, R. M. (1995). 15-Deoxy-delta 12, 14-prostaglandin J2 is a ligand for the adipocyte determination factor PPAR gamma. *Cell*, *83*, 803-812.

[45] Kliewer, S. A., Lenhard, J. M., Willson, T. M., Patel, I., Morris, D. C. & Lehmann, J. M. (1995). A prostaglandin J2 metabolite binds peroxisome proliferator-activated

receptor gamma and promotes adipocyte differentiation. *Cell*, *83*, 813-819.
[46] Schoonjans, K., Staels, B. & Auwerx, J. (1996). The peroxisome proliferator activated receptors (PPARS) and their effects on lipid metabolism and adipocyte differentiation. *Biochim. Biophys. Acta*, *1302*, 93-109.
[47] Blanquart, C., Barbier, O., Fruchart, J. C., Staels, B. & Glineur, C. (2003). Peroxisome proliferator-activated receptors: regulation of transcriptional activities and roles in inflammation. *J. Steroid Biochem. Mol. Biol.*, *85*, 267-273.
[48] Ricote, M., Li, A. C., Willson, T. M., Kelly, C. J. & Glass, C. K. (1998). The peroxisome proliferator-activated receptor-γ is a negative regulator of macrophage activation. *Nature*, *391*, 79-82.
[49] Wayman, N. S., Hattori, Y., McDonald, M. C., Mota-Filipe, H., Cuzzocrea, S., Pisano, B., Chatterjee, P. K. & Thiemermann, C. (2002). Ligands of the peroxisome proliferator-activated receptors (PPAR-gamma and PPAR-alpha) reduce myocardial infarct size. *FASEB J.*, *16*, 1027-1040.
[50] Li, M., Pascual, G. & Glass, C. K. (2000). Peroxisome proliferator-activated receptor gamma-dependent repression of the inducible nitric oxide synthase gene. *Mol. Cell. Biol.*, *20*, 4699-4707.
[51] Delerive, P., Gervois, P., Fruchart, J. C. & Staels, B. (2000). Induction of IkappaBalpha expression as a mechanism contributing to the anti-inflammatory activities of peroxisome proliferator-activated receptor-alpha activators. *J. Biol. Chem.*, *275*, 36703-36707.
[52] Delerive, P., Bosscher, K. D., Besnard, S., Berghe, W. V., Peter, J. M., Gonzalez, F. J., Fruchart, J.-C., Tedgui, A., Haegeman, G. & Staels, B. (1999). Peroxisome proliferator-activated receptor α negatively regulates the vascular inflammatory gene response by negative cross-talk with transcription factors NF-κB and AP-1. *J. Biol. Chem.*, *274*, 32048-32054.
[53] Mendez, M. & LaPointe, M. C. (2003). PPARgamma inhibition of cyclooxygenase-2, PGE2 synthase, and inducible nitric oxide synthase in cardiac myocytes. *Hypertension*, *42*, 844-850.
[54] Fahmi, H., Di Battista, J. A., Pelletier, J. P., Mineau, F., Ranger, P. & Martel-Pelletier, J. (2001). Peroxisome proliferator-activated receptor gamma activators inhibit interleukin-1beta-induced nitric oxide and matrix metalloproteinase 13 production in human chondrocytes. *Arthritis Rheum.*, *44*, 595-607.
[55] Hortelano, S., Castrillo, A., Alvarez, A. M. & Boscá, L. (2000). Contribution of cyclopentenone prostaglandins to the resolution of inflammation through the potentiation of apoptosis in activated macrophages. *J. Immunol.*, *165*, 6525-6531.
[56] Alvarez-Maqueda, M., El Bekay, R., Alba, G., Monteseirin, J., Chacon, P., Vega, A., Martin-Nieto, J., Bedoya, F., Pintado, E. & Sobrino, F. (2004). 15-deoxy-delta 12,14-prostaglandin J2 induces heme oxygenase-1 gene expression in a reactive oxygen species-dependent manner in human lymphocytes. *J. Biol. Chem.*, *279*, 21929-21937.
[57] Noyori, R. & Suzuki, M. (1993). Organic synthesis of prostaglandins: advancing biology. *Science*, *259*, 44-45.
[58] Atsmon, J., Sweetman, B. J., Baertschi, S. W., Harris, T. M. & L. Jackson Roberts, I. (1990). Formation of thiol conjugates of 9-deoxy-Δ^9,Δ^{12}(*E*)-prostaglandin D_2 and Δ^{12}(*E*)-prostaglandin D_2. *Biochemistry*, *29*, 3760-3765.

[59] Pérez-Sala, D., Cernuda-Morollón, E. & Cañada, F. J. (2003). Molecular basis for the inhibition of AP-1 DNA binding by 15-deoxy-$\Delta^{12,14}$-prostaglandin J_2. *J. Biol. Chem.*, *278*, 51251-51260.

[60] Stamatakis, K., Sánchez-Gómez, F. J. & Pérez-Sala, D. (2004). Protein modification by cyclopentenone prostaglandin addition: biological actions and therapeutic implications. *Gene Ther. Mol. Biol.*, *8*, 241-258.

[61] Straus, D. S., Pascual, G., Li, M., Welch, J. S., Ricote, M., Hsiang, C. H., Sengchanthalangsy, L. L., Ghosh, G. & Glass, C. K. (2000). 15-deoxy-delta 12,14-prostaglandin J2 inhibits multiple steps in the NF-kappaB signaling pathway. *Proc. Natl. Acad. Sci. U S A*, *97*, 4844-4849.

[62] Cernuda-Morollón, E., Pineda-Molina, E., Cañada, F. J. & Pérez-Sala, D. (2001). 15-Deoxy-$\Delta^{12,14}$-prostaglandin J_2 inhibition of NF-κB DNA binding through covalent modification of the p50 subunit. *J. Biol. Chem.*, *276*, 35530-35536.

[63] Parker, J. (1995). Prostaglandin A_2 protein interactions and inhibition of cellular proliferation. *Prostaglandins*, *50*, 359-375.

[64] Cernuda-Morollón, E., Rodríguez-Pascual, F., Klatt, P., Lamas, S. & Pérez-Sala, D. (2002). PPAR agonists amplify iNOS expression while inhibiting NF-κB: implications for mesangial cell activation by cytokines. *J. Am. Soc. Nephrol.*, *13*, 2223-2231.

[65] Appleby, S. B., Ristimaki, A., Neilson, K., Narko, K. & Hla, T. (1994). Structure of the human cyclo-oxygenase-2 gene. *Biochem. J.*, *302*, 723-727.

[66] Venema, R. C., Nishida, K., Alexander, R. W., Harrison, D. G. & Murphy, T. J. (1994). Organization of the bovine gene encoding the endothelial nitric oxide synthase. *Biochim. Biophys. Acta*, *1218*, 413-420.

[67] Pérez-Sala, D., Cernuda-Morollón, E., Pineda-Molina, E. & Cañada, F. J. (2002). Contribution of covalent protein modification to the anti-inflammatory effects of cyclopentenone prostaglandins. *Ann. N. Y. Acad. Sci.*, *973*, 533-536.

[68] Sawano, H., Haneda, M., Sugimoto, T., Inoki, K., Koya, D. & Kikkawa, R. (2002). 15-Deoxy-Delta12,14-prostaglandin J2 inhibits IL-1beta-induced cyclooxygenase-2 expression in mesangial cells. *Kidney Int.*, *61*, 1957-1967.

[69] Chatterjee, P. K., Patel, N. S., Cuzzocrea, S., Brown, P. A., Stewart, K. N., Mota-Filipe, H., Britti, D., Eberhardt, W., Pfeilschifter, J. & Thiemermann, C. (2004). The cyclopentenone prostaglandin 15-deoxy-Delta(12,14)-prostaglandin J(2) ameliorates ischemic acute renal failure. *Cardiovasc. Res.*, *61*, 630-643.

[70] Ianaro, A., Maffia, P., Cuzzocrea, S., Mazzon, E., Santoro, M. G., Di Rosa, M. & Ialenti, A. (2003). 2-Cyclopenten-1-one and prostaglandin J2 reduce restenosis after balloon angioplasty in rats: role of NF-kappaB. *FEBS Lett.*, *553*, 21-27.

[71] Ianaro, A., Ialenti, A., Maffia, P., Pisano, B. & Di Rosa, M. (2001). Role of cyclopentenone prostaglandins in rat carrageenin pleurisy. *FEBS Lett.*, *508*, 61-66.

[72] Castrillo, A., Diaz-Guerra, M. J., Hortelano, S., Martin-Sanz, P. & Bosca, L. (2000). Inhibition of IkappaB kinase and IkappaB phosphorylation by 15-deoxy- Delta(12,14)-prostaglandin J(2) in activated murine macrophages. *Mol. Cell. Biol.*, *20*, 1692-1698.

[73] Rossi, A., Kapahi, P., Natoli, G., Takahashi, T., Chen, Y., Karin, M. & Santoro, M. G. (2000). Anti-inflammatory cyclopentenone prostaglandins are direct inhibitors of IκB kinase. *Nature*, *403*, 103-108.

[74] Castrillo, A., Traves, P. G., Martin-Sanz, P., Parkinson, S., Parker, P. J. & Bosca, L.

(2003). Potentiation of protein kinase C zeta activity by 15-deoxy-delta(12,14)-prostaglandin J(2) induces an imbalance between mitogen-activated protein kinases and NF-kappaB that promotes apoptosis in macrophages. *Mol. Cell. Biol.*, *23*, 1196-1208.

[75] Moos, P. J., Edes, K., Cassidy, P., Massuda, E. & Fitzpatrick, F. A. (2003). Electrophilic prostaglandins and lipid aldehydes repress redox-sensitive transcription factors p53 and hypoxia-inducible factor by impairing the selenoprotein thioredoxin reductase. *J. Biol. Chem.*, *278*, 745-750.

[76] Mullally, J. E., Moos, P. J., Edes, K. & Fitzpatrick, F. A. (2001). Cyclopentenone prostaglandins of the J series inhibit the ubiquitin isopeptidase activity of the proteasome pathway. *J. Biol. Chem.*, *276*, 30366-30373.

[77] Shibata, T., Yamada, T., Kondo, M., Tanahashi, N., Tanaka, K., Nakamura, H., Masutani, H., Yodoi, J. & Uchida, K. (2003). An endogenous electrophile that modulates the regulatory mechanism of protein turnover: inhibitory effects of 15-deoxy-Delta 12,14-prostaglandin J2 on proteasome. *Biochemistry*, *42*, 13960-13968.

[78] Levonen, A. L., Landar, A., Ramachandran, A., Ceaser, E. K., Dickinson, D. A., Zanoni, G., Morrow, J. D. & Darley-Usmar, V. M. (2004). Cellular mechanisms of redox cell signaling: the role of cysteine modification in controlling antioxidant defenses in response to electrophilic lipid oxidation products. *Biochem. J.*, *378*, 373-382.

[79] Itoh, K., Mochizuki, M., Ishii, Y., Ishii, T., Shibata, T., Kawamoto, Y., Kelly, V., Sekizawa, K., Uchida, K. & Yamamoto, M. (2004). Transcription Factor Nrf2 Regulates Inflammation by Mediating the Effect of 15-Deoxy-Delta(12,14)-Prostaglandin J(2). *Mol. Cell. Biol.*, *24*, 36-45.

[80] Shibata, T., Yamada, T., Ishii, T., Kumazawa, S., Nakamura, H., Masutani, H., Yodoi, J. & Uchida, K. (2003). Thioredoxin as a molecular target of cyclopentenone prostaglandins. *J. Biol. Chem.*, *278*, 26046-26054.

[81] Lee, J. H., Koo, T. H., Hwang, B. Y. & Lee, J. J. (2002). Kaurane diterpene, kamebakaurin, inhibits NF-kappaB by directly targeting the DNA-binding activity of p50 and blocks the expression of antiapoptotic NF-kappaB target genes. *J. Biol. Chem.*, *277*, 18411-18420.

[82] Perry, N. B., Albertson, G. D., Blunt, J. W., Cole, A. L., Munro, M. H. & Walker, J. R. (1991). 4-Hydroxy-2-cyclopentenone: an anti-Pseudomonas and cytotoxic component from Passiflora tetrandra. *Planta Med.*, *57*, 129-131.

[83] Verbitski, S. M., Mullally, J. E., Fitzpatrick, F. A. & Ireland, C. M. (2004). Punaglandins, chlorinated prostaglandins, function as potent Michael receptors to inhibit ubiquitin isopeptidase activity. *J. Med. Chem.*, *47*, 2062-2070.

[84] Fenaille, F., Tabet, J.-C. & Guy, P. A. (2004). Identification of 4-hydroxy-2-nonenal-modified peptides within unfractionated digests using matrix-assisted laser desorption/ionization time-of-flight mass spectrometry. *Anal. Chem.*, *76*, 867-873.

[85] Burcham, P. C., Fontaine, F. R., Kaminskas, L. M., Petersen, D. R. & Pyke, S. M. (2004). Protein adduct-trapping by hydrazinophthalazine drugs: mechanisms of cytoprotection against acrolein-mediated toxicity. *Mol. Pharmacol*, *65*, 655-664.

[86] Mahon, T. M. & O'Neill, L. A. (1995). Studies into the effect of the tyrosine kinase inhibitor herbimycin A on NF-kappaB activation in T lymphocytes. Evidence for covalent modification of the p50 subunit. *J. Biol. Chem.*, *270*, 28557-28564.

[87] Ogino, S., Tsuruma, K., Uehara, T. & Nomura, Y. (2004). Herbimycin A abrogates

Nuclear Factor-kappaB activation by interacting preferentially with the IkappaB Kinase beta subunit. *Mol. Pharmacol., 65*, 1344-1351.

[88] Peng, Z., Geh, E., Chen, L., Meng, Q., Fan, Y., Sartor, M., Shertzer, H. G., Liu, Z. G., Puga, A. & Xia, Y. (2010). IkappaB kinase-beta regulates redox homeostasis by controlling the constitutive levels of glutathione. *Mol Pharmacol., 77*, 784-792.

[89] Stamatakis, K., Sánchez-Gómez, F. J. & Pérez-Sala, D. (2006). Identification of novel protein targets for modification by 15-deoxy-$\Delta^{12,14}$-prostaglandin J_2 in mesangial cells reveals multiple interactions with the cytoskeleton. *J. Am. Soc. Nephrol., 17*, 89-98.

[90] Sánchez-Gómez, F. J., Gayarre, J., Avellano, M. I. & Pérez-Sala, D. (2007). Direct evidence for the covalent modification of glutathione-S-transferase P1-1 by electrophilic prostaglandins: implications for enzyme inactivation and cell survival. *Arch. Biochem. Biophys., 457*, 150-159.

[91] Garzón, B., Gayarre, J., Gharbi, S., Díez-Dacal, B., Sánchez-Gómez, F. J., Timms, J. F. & Pérez-Sala, D. (2010). A biotinylated analog of the antiproliferative prostaglandin A_1 allows assessment of PPAR-independent effects and identification of novel cellular targets for covalent modification. *Chem. Biol. Interact., 183*, 212-221.

[92] Landar, A., Shiva, S., Levonen, A. L., Oh, J. Y., Zaragoza, C., Johnson, M. S. & Darley-Usmar, V. M. (2006). Induction of the permeability transition and cytochrome c release by 15-deoxy prostaglandin J2 in mitochondria. *Biochem. J., 394*, 185-195.

[93] Aldini, G., Carini, M., Vistoli, G., Shibata, T., Kusano, Y., Gamberoni, L., Dalle-Donne, I., Milzani, A. & Uchida, K. (2007). Identification of actin as a 15-Deoxy-Delta(12,14)-prostaglandin J(2) target in neuroblastoma cells: mass spectrometric, computational, and functional approaches to investigate the effect on cytoskeletal derangement. *Biochemistry, 46*, 2707-2718.

[94] Renedo, M., Gayarre, J., García-Domínguez, C. A., Pérez-Rodríguez, A., Prieto, A., Cañada, F. J., Rojas, J. M. & Pérez-Sala, D. (2007) Modification and activation of Ras proteins by electrophilic prostanoids with different structure are site-selective. *Biochemistry, 46*, 6607-6616.

[95] Levonen, A. L., Dickinson, D. A., Moellering, D. R., Mulcahy, R. T., Forman, H. J. & Darley-Usmar, V. M. (2001) Biphasic effects of 15-deoxy-delta(12,14)-prostaglandin J(2) on glutathione induction and apoptosis in human endothelial cells. *Arterioscler. Thromb. Vasc. Biol., 21*, 1846-1851.

[96] Gayarre, J., Stamatakis, K., Renedo, M. & Pérez-Sala, D. (2005) Differential selectivity of protein modification by the cyclopentenone prostaglandins PGA_1 and 15-deoxy-$\Delta^{12,14}$-PGJ_2: role of glutathione. *FEBS Lett., 579*, 5803-5808.

In: Prostaglandins: Biochemistry, Functions… ISBN: 978-1-61668-272-9
Editors: Gillian M. Goodwin, pp.167-186

Chapter 11

EFFECTS OF PROSTAGLANDINS AND PROSTAMIDES ANALOGUES ON OCULAR BLOOD FLOW*

Michele Vetrugno, Nicola Cardascia and Carlo Sborgia
Department of Ophthalmology, University of Bari, Italy

ABSTRACT

There have been frequent suggestions in recent years that therapeutic benefits, in particular the maintenance of an adequate blood supply to the retina or optic nerve head (ONH), may results from the topical application of drugs.

Prostanoid analogues have recently been introduced into clinical use for the management of increased intraocular pressure (IOP).[1.2] This class of compound is known to exert effect on vascular components.[3.4] Some endogenous prostaglandins have been shown to influence regional ocular blood flow and exhibit significant vasoactive proprieties in ophthalmic, ciliary and retinal arteries and veins. [5.6.7.8] In glaucomatous eyes, therapies that negatively impact ocular blood flow could interfere with the effective transport of oxygen and nutrients to already compromised areas of retina.

The purpose of this review is to examine in detail the “state of art” of the prostaglandin-induced interaction with ocular blood flow in experimental models and in glaucomatous patients.

ANATOMY AND PHYSIOLOGY OF OCULAR BLOOD FLOW

The anatomy and physiology of Ocular Blood Flow (OBF), including the ONH blood flow has been reviewed previously.[9.10.11.12.13] In order to enable a better appreciation of

---

* A version of this chapter was also published in *New Developments in Eye Research,* edited by Yuuki Inoue published by Nova Science Publishers, Inc. It was submitted for appropriate modifications in an effort to encourage wider dissemination of research

the vascular disorder in glaucoma, a summary of the anatomy and physiology of OBF is presented here.

Ocular circulation is complex, because of the necessity to supply different ocular structures with nutrients without interfering with the visual pathway. OBF is highly regulated in order to adapt to changing metabolic needs during changing visual function, to compensate for varying perfusion pressures and finally to keep the temperature at the back of the eye constant. [14.15.16]

Retinal vessels are supplied by the central retinal artery, which in turn is a branch of the ophthalmic artery.[17] The anatomy and physiology of the retinal circulation resembles the brain circulation with the exception that the retinal circulation has no autonomic innervation. Retinal circulation is characterized by a low level of flow and high level of oxygen extraction. The presence of endothelial tight junctions results in a blood—retinal barrier, similar to the blood—brain barrier. Circulation is autoregulated, meaning that within a certain range flow is independent of perfusion pressure. [18.19.20.21.22.23] Beside myogenic regulation, endothelial cells play a major role in local regulation[24,2526,27] Factors involved in this regulation are the partial pressure of oxygen and carbon dioxide, circulating and locally produced hormones like angiotensin-II and local metabolites like adenosine diphosphate.[28]

In contrast, choroidal circulation is characterized by very high flow and low oxygen extraction. Choroidal blood flow accounts for 85% of the total blood flow in the eye. Besides supplying the retina with nutrients, regulation of choroidal circulation seems to be important for maintaining temperature and volume in the eye.[29,30] The choroid is supplied by the posterior ciliary arteries, branching from the ophthalmic artery. The endothelium of the choriocapillaries is fenestrated. Poor autoregulation renders choroidal blood flow more dependent on perfusion pressure.[31.32.33.34] Like extraocular vessels, the choroid has a rich autonomic innervation.[35.36.37] Sympathetic nerves reach the eye from the superior cervical ganglion following the path of the blood vessels, while para sympathetic nerves reach the eye through the ocular motor nerve, facial nerve and through the ophthalmic and maxillary division of the trigeminal nerve. A number of neural transmitters are involved, like norepinephrine, acetylcholine, nitric oxide, vasoactive intestinal peptide and others.[38.39.40] Blood supply to the ONH is very unique by any standard.[41] The superficial layer of the ONH receives its blood supply via small branches of the central retinal artery. The prelaminar region, a small area anterior to the laminar cribrosa, is mainly supplied by branches from recurrent choroid arterioles and the short posterior ciliary arteries.[42.43.44.45.46] Venous drainage of the ONH is through the central retinal vein. The ONH seems to be the only part of the central nervous system which has no proper blood—brain barrier, with the capillaries lacking blood—brain barrier properties.[47] In addition some diffusion from the surrounding choroid into the ONH is possible. This makes ONH circulation especially sensitive to circulating molecules like endothelin-I and angiotensin II. The autoregulation seems to be less efficient than in the retina but better than in the choroid.[48.49.50.51]

Measurement Techniques

A variety of non-invasive techniques, based on entirely distinct physical principles of measurement, have been used in humans to identify changes in blood flow in the retinal, choroidal, and ONH systems. There is disagreement over the significance of many of the results' because of instrumental unreliability and of uncertainty over which blood vessels are implicated in the velocity changes observed.[52.53.54] An understanding of the principles of the operation of these techniques is essential to assessing the validity of their results and it is relevant to set them out. Furthermore, some of them have only recently come into use and many investigators may be unfamiliar with them.

Ultrasound Doppler Sonography (Color Doppler Imaging)

This makes use of the changes in frequency of a beam of ultrasound scattered back from the blood flowing in a vessel.[55] The beam of ultrasound is able to penetrate to invisible vessels-for example, the ophthalmic and central arteries that are deep and large. The blood flow can be displayed over a 20-color range.

Laser Doppler Velocimetry (LDV)

This measures the frequency change of light scattered back from the blood in order to estimate its flow in the larger visible vessels as well as that in the capillaries in the vessel free regions of the ONH.[56.57] Because it employs infrared light, which has a significant but limited penetration into the tissues, particularly the ONH, the depth of the capillaries being studied is often not clear.[58.59.60] It has been combined with a confocal scanning system to create a more complex instrument, generally known as the Heidelberg Retinal Flowmeter after its manufacturer, [61] which provides a two-dimensional image of the blood flow velocities, and a better depth resolution.[62]

Speckle Blur

This recently introduced method relies on the analysis of the speckle that is observed when a laser illuminates an object. The speckle becomes measurably blurred when he object, in this case the column of red blood cells, moves. It does not provide absolute flow values but only relative values in an individual subject, but it is reported to give reproducible values that correlate with more invasive measures.[63.64] The correlation with LDV measurements is rather disappointing, however.[65]

Fluorescein Angiography

A bolus of fluorescein is injected intravenously and a series of retinal arteriograms is recorded. These are analyzed to determine the time taken for the front of the fluorescent blood to travel from the arterial to the venous side of the ONH or over some other defined path. Recordings are acquired photographically or with a scanning laser ophthalmoscope. Although this is a versatile method, reservations about its accuracy have been expressed. [66.67.68].

Intraocular Pulse

It is claimed that pulsatory excursions make up the greater part of the blood flow through the choroid, hence the magnitude of the lOP pulse is its measure.[69] However, the basis for this conclusion is not clear. There does not even appear to be a consensus as to whether the elastic element responsible for the transmission of the pulse corresponds to the vessel walls of the choroid [70] or to the resilience of the sclera as in the Friedenwald model. Furthermore, although our ultimate concern should be the total blood perfusion, pulsatile and steady, through tissues at risk from anoxia, there is no reason to expect a simple relationship between this rate of perfusion and the raw pulse amplitude.

On the contrary, the pulse amplitude may be altered many times over by postural changes, such as sitting down, which would not be expected to promote ischemia in the retina.[71.72]

Fundus Pulse

This method uses an optical interference system to measure pulsatile movements of the macula and optic disk, which should allow the retinal and choroidal systems to be separated.[73] Although it is more specific in its target than the intraocular pulse method, the relationship between the amplitude of the pulse and the continuous blood flow through the vascular system remains unclear.

PROSTAGLANDINS

The prostaglandins (PGs) belong to a group of local hormones, the eicosanoids, named after their common 20-carbon skeleton. There are a number of naturally occurring PGs synthesised from arachidonic acid. The first indication that these local hormones existed was reported by Kurzrok and Lieb in 1930 [74] who found that human seminal fluid contracted uterine muscles. The presence of a substance in human seminal fluid that could contract smooth muscles was later verified by Goldblatt [75.76] and von Euler [77] von Euler described the presence of two lipidic substances in human seminal fluid, one water-soluble and one lipid-soluble. The substances were given the names prostaglandin E (PGE) and prostaglandin F (PGF) based on the assumption that they were produced by the prostate gland and on their solubility in ether and phosphate ("fosfat" in Swedish) respectively. It has later been found that the two PGs were not produced by the prostate gland but by the seminal vesicles and that PGs are released in many tissues where they act as local hormones with a wide range of biological activities. They are not metabolized locally but released into the circulation and then rapidly inactivated mainly during their passage through the lung.[78] The PGs constitute a family of substances and the naturally occurring PGs include PGF2a, PGE2, PGD2, and PGI2 (prostacyclin). TXA2 (thromboxane) is usually included among the PGs although it has a slightly different ring structure. The last letter in the abbreviation of the naturally occurring prostaglandins (D, E, and F) refer to the ring structure and the number, e.g. PGE2, to the number of unsaturated double bonds in the side-chains.[79] The various

PGs have strikingly different effects among themselves and in different tissues. In the eye the PGs have traditionally been regarded as potent miotics and as key players in the inflammatory process but it is now known that they are miotics only in some species and that the full inflammatory response involves a large number of substances, including the leukotrienes and other lipoxygenase products of arachidonic acid metabolism, several chemically unrelated peptides, and sensory nerves.

The different naturally occurring PGs act on a corresponding family of prostanoid receptors. All natural PGs have affinity to more than one of the prostanoid receptors which means that selective agonists or antagonists are needed to clarify the biological activity of the respective receptor in different tissues. At present there are only a few such selective drugs but those that have been developed and tested in the eye for their ocular hypotensive effect have demonstrated the importance of selectivity. Compounds based on PGF2a, but more selective for the FP-receptor, have a better therapeutic profile than the mother compound while some compounds based on other naturally occurring PGs have produced mainly ocular irritation and even increased IOP. PGs are not metabolised in the eye due to lack of the enzyme responsible for the first step of PG inactivation, PG dehydrogenase.[80] Instead, inactivation of intraocular PG activity is achieved by removal of PGs from the eye by outward-directed active transport systems for inorganic anions in the blood-ocular barriers.[81] During inflammation of the anterior segment of the eye the efficiency of this transport system is reduced.[82] It has been suggested that this characteristic of the transport system may explain the cystoid macular edema (CME) in aphakic eyes.[83] An inhibition of PG removal due to a low-grade post-operative inflammation might then permit PGs to diffuse through the vitreous body and reach the retina and induce vascular changes resulting in an edema. The mechanism by which prostaglandins reduce IOP is not the same among different species, perhaps due to a combination of different prostaglandin effects and differences in the anatomy of the outflow routes. Neither PGE1, PGE2, nor PGF2a has any effect on outflow through the trabecular meshwork in monkeys.[84] This is important information on the mechanism of action of PGs since the various parts of the outflow can be determined with precision in the monkey eye, unlike the human eye. The lack of effect of PGF2a on conventional outflow in the monkey eye has been conformed in several studies which have also shown that PGF2a does not reduce aqueous humor flow which strongly suggests that PGF2a reduces IOP mainly if not exclusively by increasing the uveoscleral outflow. [85.86.87.88.89] The first study of the effect on PGs on the human eye was reported by Giuffré.[90]

Latanoprost

Latanoprost (codenamePhXA41;13,14-dihydro-17-phenyl-18,19,20-trinor-PGF2a-isopropylester) is a PGF2a-isopropyl ester-analogue with a phenyl ring substituted for carbons 18-20 in the omega chain.[91] The double bond between carbons 13 and 14 has been saturated. Latanoprost is a more selective agonist for the FP-receptor than PGF2a, and it has only marginal effect on most of the other PG-receptors.[92] Topical application ofPGF2a-IE caused a marked vasodilatation in the conjunctiva. It also induced a short initial ocular hypertension.[93] The probable reason for the initial hypertensive phase is a dilatation of intraocular vessels. The fact that latanoprost retained the hypotensive effect on the eye is of course a strong indication that it has the same effect on uveoscleral outflow as PGF2a.

That this is the case has also been verified in studies on monkeys where uveoscleral flow was determined by the technique described by Bill.[94] These studies showed that latanoprost, like PGF2a, has a marked effect on uveoscleral outflow in monkeys.

There is no clinical technique for direct determination of uveoscleral flow, but from one study uveoscleral outflow was calculated from measurements of other parameters of aqueous humor dynamics at two pressure levels.[95] The results support the assumption that increased uveoscleral outflow is the main mechanism of action also in the human eye. The exact mechanism behind the increased uveoscleral flow is not known. The ciliary muscle of cynomolgus monkeys contains FP-receptor mRNA and protein,[96] and binding sites for EP2- and FP-receptors exist in human ciliary muscles, [97] although the dominant prostanoid receptor was found to be of the EP2 subclass. [98] The fact that latanoprost is almost twice as effective as the epimeric mixture PhXA34 also argues for a receptor-mediated effect. Two mechanisms have been proposed; relaxation of the ciliary muscle or changes in the extracellular matrix of the ciliary muscle. PGF2a relaxes ciliary muscle of the rhesus monkey,[99] and short-term treatment with large doses of PGF2a causes a dilatation of the intramuscular spaces of the ciliary muscle of cynomolgus monkeys,[100] but such changes have not been observed with latanoprost.[101]Experimental data suggest that latanoprost acts by remodeling the extracellular matrix in the ciliary muscle, thus increasing the flow of aqueous humor through the ciliary muscle bundles of the uveoscleral pathway.[102] The effect of latanoprost on ocular blood flow has been determined in humans with Langham pOBF tonograph. In this study population, pulsatile OBF increased with travoprost and latanoprost in the short term, but kept constant only with travoprost (figure 1).[103] This effect could be explained by its IOP lowering effect [104] and by the increase of ocular pulsatile amplitude and ocular pulsatile volume.[105] Variation in ocular perfusion was not observed in healthy subject treated with Latanoprost due to an effective vascular autoregulation.[106]Latanoprost increases nearly twofold pOBF compared to Unoprostone.[107] Anyway, studies on hemodynamic by means color Doppler imaging ultrasound of retrobulbar vessels (central retinal artery, short posterior ciliary arteries, and ophthalmic artery) did not reveal variation of blood velocity.[108.109.110.111.112] Further studies reveal clinical outcomes could be improved by latanoprost reducing IOP and increasing pOBF[113] (table 1).

Travoprost

Travoprost (isopropyl (Z)-7- [(1 R ,2 R ,3 R ,5 S)-3,5-dihydroxy-2-[(1 E ,3 R)-3-hydroxy-4-[((alpha),(alpha),(alpha)-trifluoro- m -tolyl)oxy]-1- butenyl]cyclopentyl]-5-hept-enoate), a highly selective and potent analogue of the prostaglandin PGF(2)(alpha), has recently been approved and marketed as a topical ocular hypotensive agent for the treatment of ocular hypertension and glaucoma. Following absorption into the eye, the free acid form of travoprost interacts with

Table 1. Latanoprost: Ocular Blood Flow evaluation

Author (year)	Methods	Trial	Patients	Follow up	Ocular Blood Flow
Bill A. (1965)	Aqueo variations	Experimental	Monkey Rhesus	NA	increased
Poyer J F (1995)	Ciliary Muscle	Experimental	Monkey Rhesus	NA	Relaxing (increased)
Ravinet E (2003)	Ciliary Muscle	Experimental	Monkey Rhesus	NA	Extracellular matrix (increased)
Nicolela MT (1996)	Echo color doppler	Clinical	9 POAG	7 days	Not Affected
Vetrugno M (1998)	Langham OBF	Clinical	12 POAG	180 days	increased
McKibbin M (1999)	Langham OBF	Clinical	19 NTG	28 days	increased
Kuba GB (2001)	Langham OBF	Clinical	30 HEALTY	7 days	Not Affected
Georgopoulos GT (2002)	Langham OBF	Clinical	24 POAG	7 days	increased
Geyer O (2002)	Langham OBF	Clinical	19 HEALTY	8 hours	increased
Harris A (2003)	Echo color doppler	Clinical	20 NTG	28 days	Not Affected
Inan UU (2004)	Echo color doppler	Clinical	29 OHT	5 days	increased
Akarsu C (2004)	Echo color doppler	Clinical	23 POAG / OHT	8 hours	Not Affected

The endogenous FP prostanoid receptor to enhance aqueous humor outflow and lower intraocular pressure (IOP). Travoprost is distinguished from other marketed prostaglandin analogues in that it is a full agonist at the prostaglandin receptor. It is also highly selective with little or no affinity for other prostanoid or non-prostanoid receptors in the eye.[114] In isolated porcine ciliary arteries appears to have no vasoconstrictive properties in isolated porcine ciliary arteries.[115] In humans travoprost improves ocular blood flow, reducing IOP and balancing choroidal capillary autoregulation (figure 1, table 2).[103]

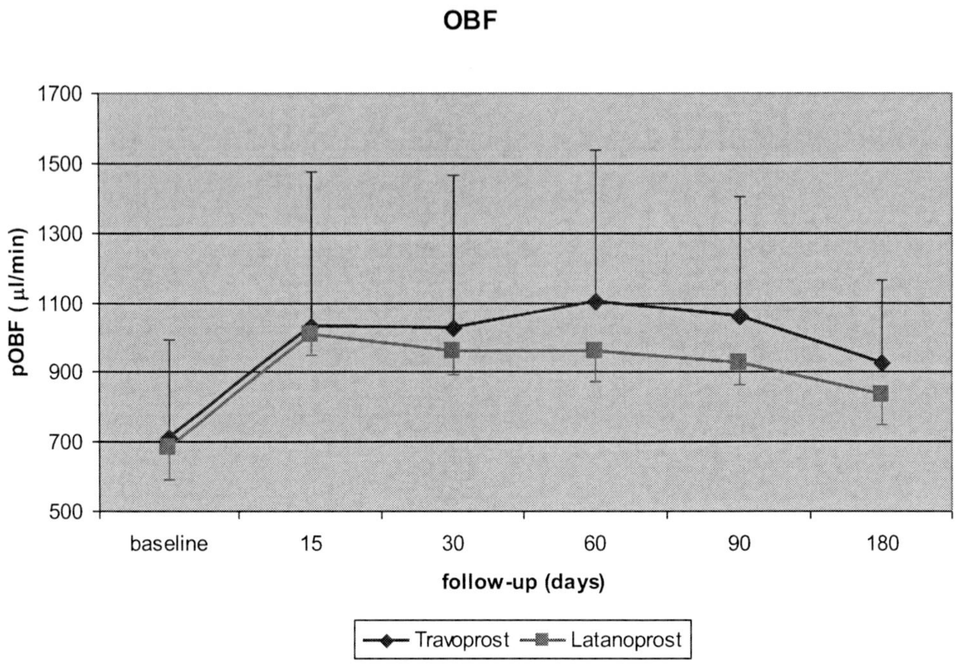

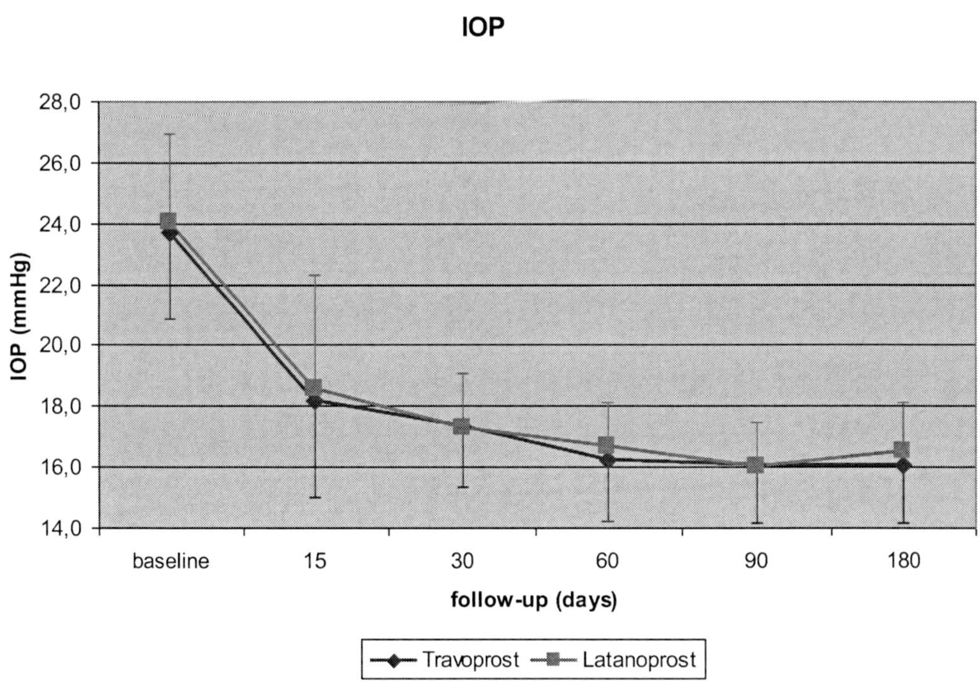

Figure 1. Variations of IOP and pOBF in patients affected by POAG treated with topical latanoprost and topical travoprost.

Table 2. Travoprost: Ocular Blood Flow evaluation

Author (year)	Methods	Trial	Patients	Follow up	Ocular Blood Flow
Allemann R (2003)	Myography ciliary arteries	Experimental	porcine	NA	Not Affected
Inan UU (2004)	Echo color doppler	Clinical	19 HEALTY	5 days	increased
Cardascia N (2004)	Langham OBF	Clinical	16 POAG	180 days	increased

Unoprostone

Isopropyl 20-ethyl-9a,11a-dihydroxy-15-keto-cis-D5-prostanoate (unoprostone, UF-021, Rescula) is another PGF2a-related compound. It has been on the market in Japan for more than one year, but there is little experience with the drug outside Japan. It was reported that 0.12% solution reduced IOP in normal eyes without any effect on aqueous flow, outflow facility, or episcleral venous pressure[116] By means of a laser speckle tissue circulation analyzer was measured the normalized blur, a quantitative index of blood flow velocity, in the optic nerve head (ONH) and choroid-retina revealing a reduction in vascular resistance.[117.118] (table 3).

Table 3. Unoprostone: Ocular Blood Flow evaluation

Author (year)	Methods	Trial	Patients	Follow up	Ocular Blood Flow
Sakurai M (1991)	Aqueo variations	Experimental	NA	NA	Not Affected
Kojima S (1997)	laser speckle analyzer	Clinical	9 HEALTY	6 hours	increased
Sponsel WE (2002)	Langham OBF	Clinical	25 POAG	28 days	increased
Makimoto Y (2002)	laser speckle analyzer	Clinical	11 HEALTY	21 days	increased

Prostamides

Prostamides are the latest member of the fatty acid amide family, which include anadamide,[119] oleamide, [120] and palmitylethanolamide [121]. Those are potent hypotensive agents. They are naturally occurring in tissues [122], where they are biosynthesized from anandamide exclusively by COX-2 and an aldoketoreductase, cyclooxygenase 1 (COX-1) in not involved.[123] Model experiments showed exogenously administered anantamide is converted to prostamide by lung tissue. The role of prostamides in the tissue in which they occur remains to be elucidated. Nevertheless, on contemplating the possible role of prostamides in the eye, it is of interest to note that prostamides lower IOP, COX-2 has been reported as constitutively present in the human eye [124], COX-2 mutation have been reported in primary open angle glaucoma patients [125].

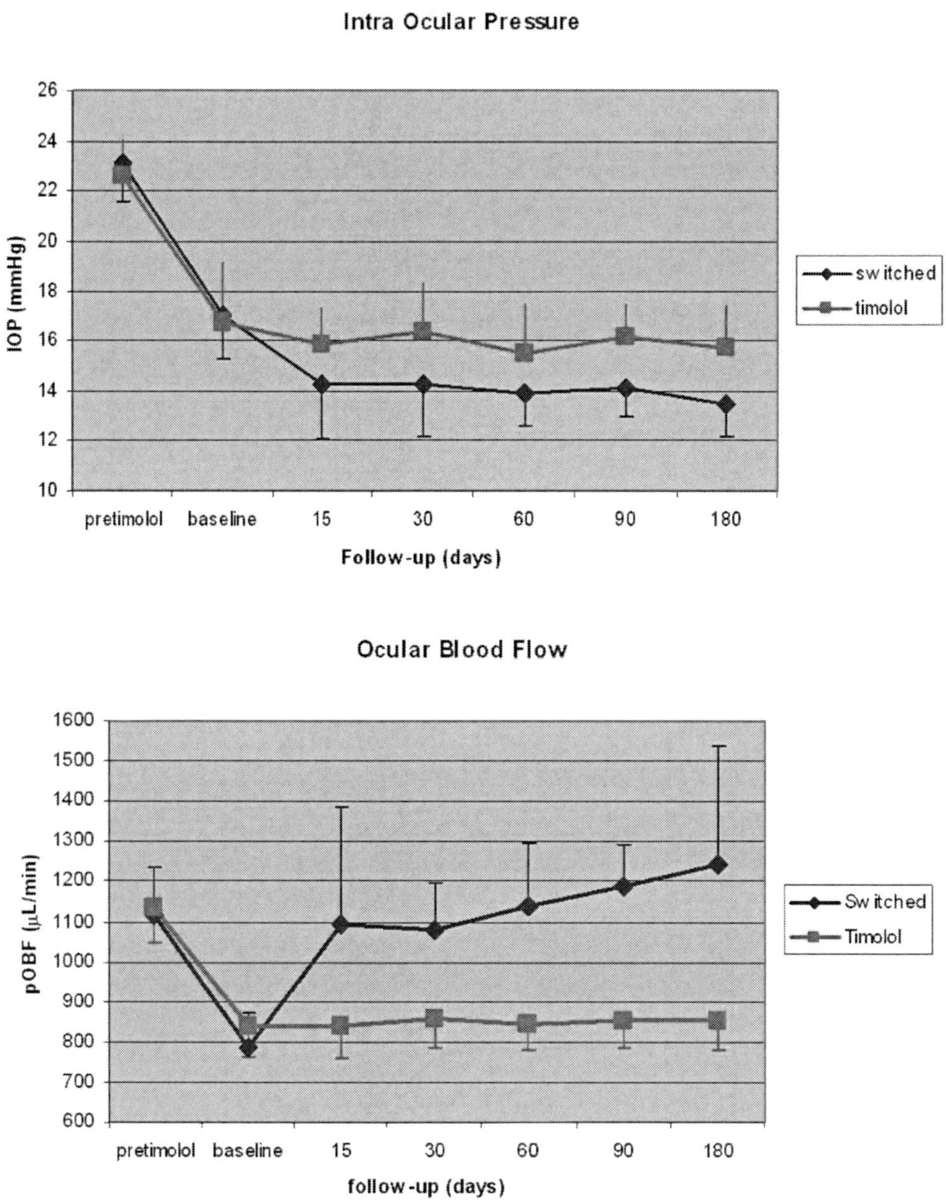

Figure 2. Variations of IOP and pOBF in patients affected by POAG switched from topical timolol to topical bimatoprost.

Bimatoprost

Bimatoprost is the ethyl amide derivative of 17-phenyl-trinor prostaglandin F(2alpha). It is a new ocular hypotensive agent that lowers intraocular pressure (IOP) in normal, ocular hypertensive, and glaucomatous eyes. Its mechanism of action has been studied in normal human subjects. Bimatoprost mildly stimulates the rate of aqueous humor flow during the day (13%) and at night (14%). Its ocular hypotensive action is due primarily to a 26% reduction in the tonographic resistance to outflow. Thus, bimatoprost enhances the pressure-sensitive outflow pathway. Additional beneficial effects may include an increase in the rate of flow via the pressure-insensitive outflow pathway (sometimes called the "uveoscleral outflow pathway") and a lowering of the extraocular recipient pressure (sometimes called "episcleral venous pressure").

Reduction of tonographic resistance to aqueous humor outflow reduces steady-state IOP, an effect that is beneficial for the treatment of glaucoma.[126] It appears to have, in vitro, some vasoactive properties in porcine ciliary arteries. [127] In humans, it does not affect retinal arteriolar calibre. [128] Unpublished investigations show it efficacy on ocular blood flow in humans: after 6 months pOBF increases of 48% from the baseline.[129] Upcoming studies reveals Bimatoprost achieves improvement of 29% of choroidal blood flow in humans after 6 months of therapy, not influenced by IOP variations (figure 2).[130] Improvement of ocular blood flow and IOP reduction demonstrate clinical improvements in patients affected by normal tension glaucoma[131.] (table 4).

Table 4. Bimatoprost: Ocular Blood Flow evaluation

Author (year)	Methods	Trial	Patients	Follow up	Ocular Blood Flow
Joyner WL (1986)	Myography retinal arteries	Experimental	Retinal arteries	NA	Not Affected
Allemann R (2003)	Myography ciliary arteries	Experimental	porcine	NA	increased
Brubaker RF (2001)	Aqueo variations	Clinical	25 POAG	3 days	increased
Agarwal HC (2003)	Langham OBF	Clinical	32 POAG	3 months	increased
Akarsu C (2004)	Echo Color Doppler	Clinical	26 POAG OHT	1 month	Not Affected
Inan UU (2004)	Echo Color Doppler	Clinical	19 HEALTY	6 hours	increased
Zeitz O (2004)	Echo Color Doppler	Clinical	9 NTG	3-5 weeks	Not Affected
Vetrugno M (2004)	Langham OBF	Clinical	36 POAG	180 days	increased

Discussion

Endogenous prostaglandins have been shown to mediate microvascular responses in a variety of tissues, including those of the eye, but naturally occurring prostanoid also typically possess rather modest selectivity for their respective native receptor subtypes and often exhibits activity at other prostaglandin receptor.[132]

Several investigations reported an increase or preservation of ocular blood flow but the precise mechanism remain unveiled. The activity of prostaglandins on ocular blood flow is twofold: direct activity on intraocular blood vessels; indirect activity mediated by IOP lowering. Considering the direct pathway, PGs do not act on β or α receptors of ocular vessels but they probably act on blood flow playing a role in the inflammatory pathway. This issue, anyway, is to be confirmed in further investigations. The indirect mechanism refers to the action on IOP. The reduction of pressure on the vessels improves ocular blood flow.

We have to believe there is a synergic action of all the factors previously described that hesitate on blood flow increasing. Prostaglandins and Prostamides represent the ultimate strategies for the treatment of glaucoma and they are the only molecules that have a direct or indirect action on ocular blood flow. We expect an intensive search for improved alternatives as there is no obvious reason to believe that this first generation will turn out to be the ultimate prostaglandin for the treatment of glaucoma.

References

[1] Yamamoto, T., Kitazawa, Y., Azuma, I. & Masuda, K. (1997). Clinical evaluation of UF-021 (Rescula; isopropyl unoprostone). *Surv. Ophthalmol., 41*, S99-S103

[2] Bito, L. Z. (2001). A new approach to the medical management of glaucoma, from the bench to the clinic, and beyond. *Invest. Ophthalmol. Vis Sci., 42*, 119-28

[3] Kennedy, I., Coleman, R. A., Humprey, P. P. A., Levy, P. & Lumley, P. (1982). Studies on the characterization of Prostanoid receptors: a proposed classification. *Prostaglandins, 24*, 667-89

[4] Giles, H. & Leff, P. (1988). The biology and pharmacology of PGD_2. *Prostaglandins, 35*, 277-300

[5] Nielsen, P. & Nyborg, N. C. B. (1989). Calcium antagonist-induced relaxation of the prostaglandin-$F_{2\alpha}$ response of isolated calf retinal resistance arteries. *Exp. Eye Res., 48*, 329-35.

[6] Nielsen, P. & Nyborg, N. C. B. (1990). Contractile e relaxing effects of arachidonic acids derivates on isilated bovine retinal resistance arteries. *Exp. Eye Res., 50*, 305-11.

[7] Kulkarni, P. S. (1995). Prostaglandin F_2□ vasoconstriction of bovine retinal microcirculation is via thromboxane receptors. *Invest. Ophthalmol. VIs. Sci., 36*, S137.

[8] Astin, M. (1998). Effect of prostaglandin E_2 ,F_2□ and latanoprostacid on isolated ocular blood vessel in vitro. *J. Ocul. Pharmacol. Ther., 14*, 119-28.

[9] Flammer, J. & Orgül, S. (1998). Optic nerve blood-flow abnormalities in glaucoma. *Prog. Retin. Eye Res., 17*, 267-289.

[10] Hayreh, S. S. (1996). Blood supply of the optic nerve head. *Ophthalmologica, 210*, 285-295.

[11] Bill, A. & Nilsson, S. F. (1985). Control of ocular blood flow. *J. Cardiovasc. Pharmacol., 7*, (Suppl. 3), S96-S102.

[12] Bill, A. & Nilsson, S. F. (1985). Control of ocular blood flow. *J. Cardiovasc. Pharmacol., 7*, (Suppl. 3), S96-S102.

[13] Buechi, E. R. (1995). The blood supply to the optic nerve head. In: Kaiser, H.J., Flammer, J., Hendrickson, P. (Eds.), *Ocular Blood Flow, Glaucoma-Meeting*. Karger, Basel, 1-8.

[14] Alm, A. & Bill, A. (1973). Ocular and optic nerve blood flow at normal and increased intraocular pressures in monkeys (Macaca irus): a study with radioactively labeled microspheres including flow determinations in brain and some other tissues. *Exp. Eye Res., 15*, 15-29.

[15] Delaey, C.. Van Dc Voorde, J. (2000). Regulatory mechanisms in the retinal and choroidal circulation. *Ophthalmic Res., 32*, 249-256.

[16] Alm, A. (1998). Optic nerve and choroidal circulation: physiology. In: Haefliger, 1.0., Flammer, J. (Eds.), *Nitric Oxide and Endothelin-l in the Pathogenesis of Glaucoma.* Lippincott-Raven Publishers, *Philadelphia*, 34-43.

[17] Olver, J. M. (1998). Angioarchitecture of the human optic nerve. In: C., Bisantis, G. Carella, (Eds.), *Vascular System of the Optic Nerve.* Innovation-News-Communication, Rome, 29-44.

[18] Grunwald. I. E.. Sinclair, S. I. I. & Riva, C. E. (1982). Autoregulation of the retinal circulation in response to decrease of intraocular pressure below normal. *Invest. Ophthalmol. Vis. Sci., 23,* 124-127.

[19] Pournaras, C. J. & Riva, C. E. (2000). Studies of the hemodynamics of the optic nerve head using laser Doppler flowmetry. *J. Fr. Ophthalmol., 24*, 199-205.

[20] Riva, C. E., Sinclair, S. H. & Grunwald, J. E. (1981). Autoregulation of retinal circulation in response to decrease of perfusion pressure. *Invest. Ophthalmol. Vis. Sci., 21*, 34-38.

[21] Dumskyj, M. J., Eriksen, J. E., Dore, C. J. & Kohner, E. M. (1996). Autoregulation in the human retinal circulation: assessment using isometric exercise, laser Dopplcr velocimetry, and computer- assisted image analysis. *Microvasc. Res., 51*, 378-392.

[22] Rassam, S. M., Patel, V., Chcn, H. C., Kohner, E. M. (1996). Regional retinal blood flow and vascular autoregulation. *Eye 10, 33*, 1-337.

[23] Schulte, K., Wolf, S., Arend, 0., Harris, A., Henle, C. & Reim, M. (1996). Retinal hemodynamics during increased intraocular pressure. *Ger. J. Ophthalmol., 5*, 1-5.

[24] Haefliger, I. O., Flammer, J. & Luscher, T. F. (1992). Nitric oxide and endothelin-l-l are important regulators of human ophthalmic artery. *Invest. Ophthalnsol. Vis. Sci., 33*, 2340-2343.

[25] Meyer, P., Flammer, J. & Luschcr, T. F. (1995). Local action of the renin angiotensin system in the porcine ophthalmic circulation: effects of ace-inhibitors and angiotensin receptor antagonists. *Invest. Ophthalmol. Vis. Sci., 36*, 555-562.

[26] Haefliger. 1. 0., Flammer, J., Bény, J. L. & LOscher, T. (2001). Endotheliuni-dependent vasoactive modulation in the ophthalmic circulation. *Prog. Retin. Eye Res., 20*, 209-225.

[27] Orgul, S., Gugleta, K. & Flammer, J. (1999). Physiology of perfusion as it relates to the optic nerve head. *Surv. Ophthalmol., 43*, Sl7-S26.

[28] Stefánsson, E., Wagner, H. G. & Seida, M. (1988). Retinal blood flow and its

autorcgulation measured by intraocular hydrogen clearance. *Exp. Eye Res., 47*, 669-678.

[29] Flugel, C., Tamm, E. R., Mayer, B. & Lutjen-Drecoll, F. (1994). Species differences in ehoroidal vasodilative innervation: evidence for specific intrinsic nitrergie and vip-positive neurons in the human eye. *Invest. Ophthalmol. Vis. Sci., 35*, 592-599.

[30] Flugel-Koch, C., May, C. A. & Lutjen-Drecoll, E. (1996). Presence of a contractile cell network in the human choroid. *Ophtha, 210*, 296-302.

[31] Delaey, C. & Van Dc Voorde, J. (2000). Regulatory mechanisms in the retinal and choroidal circulation. *Ophthalmic Res. 32*, 249-256.

[32] Geiser, M. H., Riva, C. E., Dorner, G. T., Diermann, U., Luksch. A. & Schmetterer, L. (2000). Response of ehoroidal blood flow in the foveal region to hyperoxia and hyperoxia-hypercapnia. *Curr. Eye Res., 21*, 669-676.

[33] Findl, O., Rainer, G., Dallinger, S., Dorner, G., Polak, K., Kiss. B., Georgopoulos, M., Vass, C. & Schmetterer, L. (2000). Assessment of optic disk blood flow in patients with open-angle glaucoma. *Am. J. Ophthalmol., 130*, 589-596.

[34] Riva, C. E., Titze, P., Hero, M. & Petrig, B. L. (1997b). Effect of acute decreases of perfusion pressure on choroidal blood flow in humans. *Invest. Ophthalmol. Vis. Sci., 38*, 1752-1760.

[35] Chou, P. I., Lu, D. W. & Chen, J. T. (2000). Bilateral superior cervical ganglionectomy increases choroidal blood flow in the rabbit. *Ophthalmologica, 214*, 421-425.

[36] May, C. A., Hayreh, S. S., Ossoining, K., Kaufman, P. L. & Lütjen Drecoll, E. (1997). Choroidal ganglion cell plexus and retinal vasculature in laser-induced monkey glaucoma. *Ophthalmologica, 211*, 161-171.

[37] Steinle, J. J., Krizsan-Agbas, D. & Smith, P. G. (2000). Regional regula tion of choroidal blood flow by autonomic innervation in the rat. *Am. J. Physiol. Regul. Integr. Comp. Physiol., 279*, R202-R209.392.

[38] Haelliger, I. O., Dettmann, E. S., Flammer, J. (1999a). Update on nitric oxide and endothehn-l in glaucoma. In: L. E., Pillunat, A., Harris, D. R., Anderson, E. L. Greve, (Eds.). *Current Concepts on Ocular Blood Flow in Glaucoma.* Kugler Publications, The Hague, 197-204

[39] Koss, M. C. (1999). Functional role of nitric oxide in regulation of ocular blood flow. *Fur. J. Pharmacol., 374*, 16 1-174.

[40] Schrodl, F., Schweigert, M., Brehmer, A. & Neuhuber, W. L. (2001). Intrinsic neurons in the duck choroid are contacted by CGRP immunoreactive nerve fibres: evidence for a local pre-central reflex arc in the eye. *Exp. Eye Res., 72*, 137-146.

[41] Olver, J. M. (1998). Angioarchitecture of the human optic nerve. In: C., Bisantis, G. Carella, (Eds.), *Vascular System of the Optic Nerve.* Innovation-News-Communication, Rome, 29-44.

[42] Hayreh, S. S. (1996). Blood supply of the optic nerve head. *Ophthal mologica, 210*, 285-295.

[43] Hayreh. S. S. (2001). The blood supply of the optic nerve head and the evaluation of it myth and reality. *Prog. Retin. Eye Res., 20*, 563-593.

[44] Ernest, J. T. (1976). Editiorial: vasculature of the anterior optic nerve. *Am. J. Ophthalmol., 82*, 509-510.

[45] Lieberman, M. F., Maumenee, A. E. & Green, W. R. (1976). Histologic studies of the vasculature of the anterior optic nerve. *Am. J. Ophthalmol., 82*, 405-423.

[46] Onda, E., Cioffi, G. A., Bacon, D. R. & van Buskirk, E. M. (1995). Microvasculature of the human optic nerve. *Am. J. Ophthalmol., 120*, 92-102.

[47] Hofman, P)., Hoyng, P., vanderWerf, F., Vrensen, G. F. & Schlingeniann, R. O. (2001. Lack of blood-brain barrier properties in microvessels of the prelaminar optic nerve head. *Invest. Ophthalmol. Vis. Sci., 42*, 895-901.

[48] Pournaras, C. J. & Riva, C. E. (2000). Studies of the hemodynamics of the optic nerve head using laser Doppler flowmetry. J. *Fr. Ophthalmol., 24*, 199-205.

[49] Movaffaghy, A., Chamot, S. R., Petrig, B. L. & Riva, C. E. (1998). Blood flow in the human optic nerve head during isometric exercise. *Exp. Eye Res., 67*, 561-568.

[50] Riva, C. E., Hero, M., Titze, P. & Petrig, B. (1997a). Autoregulation of human optic nerve head blood flow in response to acute changes in ocular perfusion pressure. *Graefes Arch. Clin. Exp. Ophthalmol., 235*, 618-626.

[51] Shonat, R. D., Wilson, D. F., Riva, C. E. & Cranstoun, S. D. (1992). Effect of acute increases in intraocular pressure on intravascular optic nerve head oxygen tension in cats. *Invest. Ophthalmol. Vis. Sci., 33*, 3174-3 180.

[52] Hayreh, S. S. The (1994). Von Sallman Lecture. The optic nerve head circulation in health *disease. Exp. Eye Res., 61*, 259-72, 1995.

[53] Hayreh, S. S. (1997). Factor s influencing blood flow in the optic nerve head. *J. Glaucoma., 6*, 412-25.

[54] Yaoeda, K., Shirakashi, M., Funaki, S., et al: (2000). Measurement of microcirculation in the optic nerve head by laser speckle flowgraphy and scanning laser Doppler Flowmetry. *Am. J. Ophthalmol., 129*, 734-9,

[55] Rankin, S. J. (1999). Color Doppler Imaging of the retrobulbar circulation in glaucoma. *Surv. Ophthamol., 43*, S176-82.

[56] Riva, C. E., Feke, G. T., Eberli, B., Benary, V. & Bidirectional, L. D. V. (1979). system for absolute measurement of blood speed in retinal vessels, *Applied Optics, 18*, 2301-6.

[57] Rica, C. E. (1981). Grunwald JE, Sinclair SH, O'Keefe K: Fnndus camera based retinal LDV. *Applied Optics, 20*, 1 I 7-20.

[58] Petrig, B. L., Riva, C. E. & Hayreh, S. S. (1999). Laser Doppler flowmetry and optic nerve head blood flow. *Am. J. Ophthalmol., 127*, 413-25.

[59] Piltz-Seymour, J. R. (1999). Laser Doppler flowmetry of the optic nerve head in glaucoma. *Surv. Ophthalmol., 43*, S191-8.

[60] Tomidokoro, A., Araie, M., Tamaki, Y. & Tomita, K. (1998). In vivo measurement of iridial circulation using laser speckle phenomenon. *Invest. Ophthalmol. Vis. Sci., 39*, 364-71.

[61] Haefliger, I. O., Lietz, A., Griesser, S. M., et al: (1999). Modulation of Heidelberg Retinal Flowmeter parameter flow at the papilla of healthy subjects: effect of carbogen, oxygen, high high intraocular pressure, and beta-blockers. *Surv. Ophthalmol., 43*, S59-65.

[62] Nicolela, NI'F., Hnik, P., Schulzer, M. & Drance, S. M. (1997). Reproducibility of retinal and optic nerve head blood flow measurements with scanning laser Doppler Flowmetry. *J. Glaucoma, 6*, 157-64.

[63] Sugiyama, T., Utsumi, T., Azuma, I. & Fujii, H. (1996). Measurement of Optic Nerve Head circulation: comparison of laser speckle and hydrogen clearance methods. *Jpn. J. Ophthalmol., 40*, 339-43.

[64] Tamaki, Y., Araie, M., Kawamoto, E., et al: (1995). Non-contact, two-dimensional

measurement of tissue circulation in choroid and optic nerve head using laser speckle phenomenon. *Exp. Eye Res., 60*, 373-83.

[65] oekada, K., Shirakashi, M., Funaki, S., et al: (2000). Measurements of microcirculation in the optic nerve head by laser speckle flowgraphy and scanning laser Doppler Flowmetry. *Am. J. Ophthalmol., 129*, 734-9.

[66] Flower, R. W. & Hochheimer, B. F. (1977). Quantification of indicator dye concentration in ocular blood vessel. *Exp. Eye Res., 25*, 103-11.

[67] Hayreh, S. S. (1997). Evaluation of optic nerve head circulation: review of the methods used. *J. Glaucoma., 6*, 319-30.

[68] Tomic, L., Maepea, O., Sperberg, G. O. & Alm, A. (2001). Comparison of retinal transit times and retinal blood flow: a study in monkeys. *Invest. Ophthalmol. Vis. Sci., 42*, 752-5.

[69] Langham, M. E., Farrell, R. A., O'Brien, V., et at: (1989). Blood flow in the human eye. *Acta Ophthalmol., 191*, (Suppl), 9-1 3.

[70] Flower, R. W. & Klein, G. J. (1990). Pulsatile flow in the choroidal circulation: a preliminary investigation. *Eve, 4*, 310-8.

[71] Bain, W. E. S. & Maurice, D. M. (1959). Physiological variations in the intraocular pressure. *Trans. Ophthalmol. Soc.* UK, *79*, 249-60.

[72] Maurice, D. M. (1958). A recording tonometer. *Br. J. Ophthalmol., 42*, 321-35.

[73] Schmetterer, L., Woltz, M., Lexer, F., et al: (1995). The effect of hyperoxia and hypercapnia on fundus pulsation in the macular and optic disk region in healthy young men. *Exp. Eye Res., 61*, 685-90.

[74] Kurzrok, C. & Lieb, C. (1930). Biochemical studies on human semen: action of semen on the human uterus. *Proc. Soc. Exp. Biol. Med., 28*, 268-272.

[75] Goldblatt, M. W. (1933) A depressor substance in seminal fluid. *J. Soc. Chem. Industry, 5*, 1056-1057.

[76] Goldblatt, M. W. (1935). Properties of human seminal plasma. *J. Physiol.* (London), *84*, 208-218.

[77] von Euler, U. S. (1934). Zur Kenntnis der Pharmakologischen Wirkungen von Nativsekreten und Extrakten mannlicher accessorischer Geschlechtsdrusen. *Nauyn Schmiedeberg's Arch. Exp. Pathol. Pharmacol., 175*, 78-84.

[78] Ferreira, S. H. & Vane, J. R. (1967). Prostaglandines: their disappearance from and release into the circulation. *Nature, 216*, 868-873.

[79] Coleman, R. A., Smith, W. L. & Narumiya, S. (1994). International union of pharmacology classifocation of prostanoid receptors; properties, distribution, and structure of the receptors and their subtypes. *Pharmacol. Rev., 46*, 205-229.

[80] Eakins, K. E. (1976). Prostaglandins and the eye. In *Advances in Prostaglandin Research. Prostaglandins, Physiological, Pharmacological and Pathological Aspects,* ed. S. M.M Karim, pp. 63-81. University Park Press, Baltimore.

[81] Bito, L. Z. & Salvador, E. V. (1972). Intraocular fluid dynamics. III. The site and mechanism of prostaglandin transfer across the blood intraocular fluid barriers. *Exp. Eye Res., 14*, 233-241.

[82] Bito, L. Z. (1974). The effects of experimental uveitis on anterior uveal prostaglandin transport and aqueous humor composition. *Invest. Ophthalmol., 13*, 299-306.

[83] Miyake, K., Shirasawa, E. & Hikita, M. (1989). Hypotheses on the role of prostaglandins in the pathogenesis of epinephrine maculopathy and aphakic cystoid

macular edema. In The Ocular Effects of Prostaglandins and Other Eicosanoids, ed. L. Z. *Bito and J. Stjernschantz,* 265-276. Alan R. Liss. Inc., New York.

[84] Kaufman, P. L. (1986). Effects of intracamerally infused prostaglandins on outflow facility in cynomolgus monkey eyes with intact or retrodisplaced ciliary muscle. *Exp. Eye Res., 43*, 819-827.

[85] Crawford, K. & Kaufman, P. L. (1987). Pilocarpine antagonizes prostaglandin F2a-induced ocular hypotension in monkeys. Evidence for enhancement of uveoscleral outflow by prostaglandins F2a. *Arch. Ophthalmol., 105*, 1112-1116.

[86] Crawford, K., Kaufman, P. L. & Gabelt, B. T. (1987). Effects of topical PGF2a on aqueous humour dynamics in cynomolgus monkeys. *Curr. Eye Res., 6*, 1035-1044.

[87] Nilsson, S. F. E., Samuelsson, M., Bill, A. & Stjernschantz, J. (1989). Increased uveoscleral outflow as a possible mechanism of ocular hypotension caused by prostaglandin F2a-1-isopropylester in the cynomolgus monkey. *Exp. Eye Res., 48*, 707-716.

[88] Gabelt, B. T. & Kaufman, P. L. (1989). Prostaglandin F2a increases uveoscleral outflow in the cynomolgus monkey. *Exp. Eye Res., 49*, 389-402.

[89] Gabelt, B. T. & Kaufman, P. L. (1990). The effect of prostaglandin F2a on trabecular outflow facility in cynomolgus monkeys. *Exp. Eye Res., 51*, 87-91.

[90] Giuffré , G. (1985) The effects of prostaglandin F2a in the human eye. *Graefe's Arch. Clin. Exp. Ophthalmol., 222*, 139-141.

[91] Stjernschantz, J. & Resul, B. (1992). Phenylsubstituted prostaglandin analogs for glaucoma treatment. *Drugs of the Future, 17*, 691-704.

[92] Stjernschantz, J., SeleÂ, N. G., SjoÈ quist, B. & Resul, B. (1995). Preclinical pharmacology of latanoprost, a phenylsub-stituted PGF2a analogue. In *Advances in Prostaglandin, Tromboxane, and Leukotriene Research*, ed. B. Samuelsson, Vol. 23, 513-518. Raven Press, New York.

[93] Villumsen, J. & Alm, A. (1989). Prostaglandin F2a-isopropy-lester eye drops. Effects in normal human eyes. *Br. J. Ophthalmol., 73*, 419-426.

[94] Bill, A. (1965). The aqueous humor drainage in the cynomolgus monkey (Macacairus) with evidence for unconventional routes. *Invest. Ophthalmol., 4*, 911-919.

[95] Toris, C. B., Camras, C. B. & Yablonski, M. E. (1993). Effects of PhXA41, a new prostaglandin F2a analog, on aqueous humor dynamics in human eyes.*Ophthalmology, 100*, 1297-1304.

[96] Ocklind, A., Lake, S., Wentzel, P., NisteÂ r, M. & Stjernschantz, J. (1996). Localization of the prostaglandin F2a receptor messenger RNA and protein in the cynomolgus monkey eye. *Invest. Ophthalmol. Vis. Sci., 37*, 716-726.

[97] Matsuo, T. & Cynader, M. S. (1992). Localisation of prostaglandin F2a and E2 binding sites in the human eye. Br. *J. Ophthalmol., 76*, 210-213.

[98] Matsuo, T. & Cynader, M. S. (1993). The EP2 receptor is the predominant prostanoid receptor in the human ciliary muscle. *Br. J. Ophthalmol., 77*, 110-114.

[99] Poyer, J. F., Millar, C. & Kaufman, P. L. (1995) Prostaglandin F2a effects on isolated rhesus monkey ciliary muscle. *Invest. Ophthalmol. Vis. Sci., 36*, 2461-2465.

[100] Lũtjen-Drecoll, E. & Tamm, E. (1988). Morphological study of the anterior segment of cynomolgus monkey eyes following treatment with prostaglandin F2a. *Exp. Eye Res., 47*, 761-769.

[101] Svedbergh, B. & Forsberg, I. (1993). A morphological study on the effects of chronic

administration of latanoprost (LP) on the ciliary muscle and trabecular meshwork in monkeys. *Invest. Ophthalmol. Vis. Sci., 34*, (ARVO), 932.

[102] Ravinet, E., Mermoud, A., Brignoli, R. (2003). Four years later: a clinical update on latanoprost. *Eur. J. Ophthalmol.,* Mar; *13(2)*, 162-75.

[103] Cardascia, N., Vetrugno, M., Trabucco, T., Cantatore, F. & Sborgia, C. Effects of Travoprost Eye Drops on Intraocular Pressure and Pulsatile Ocular Blood Flow: A 180-Day, Randomized, Double-Masked Comparison with Latanoprost Eye Drops in Patients with Open-Angle Glaucoma. *Cur. Therapeut. Res., 64(7)*, 389-400.

[104] Liu, C. J., Ko, Y. C., Cheng, C. Y., Chou, J. C., Hsu, W. M., Liu, J. H. (2002). Effect of latanoprost 0.005% and brimonidine tartrate 0.2% on pulsatile ocular blood flow in normal tension glaucoma. *Br. J. Ophthalmol.,* Nov; *86(11)*, 1236-9.

[105] Georgopoulos, G. T., Diestelhorst, M., Fisher, R., Ruokonen, P., Krieglstein, G. K. (2002). The short-term effect of latanoprost on intraocular pressure and pulsatile ocular blood flow. *Acta Ophthalmol. Scand.*, Feb, *80(1)*, 54-8.

[106] Kuba, G. B., Kurnatowski-Billion, M., Neuser, A. M., Austermann, P. (2001). Effect of latanoprost on ocular hemodynamics and contrast sensitivity. *Ophthalmologe.,* Jun; *98(6)*, 535-40.

[107] Sponsel, W. E., Paris, G., Trigo, Y. & Pena, M. (2002). Comparative effects of latanoprost (Xalatan) and unoprostone (Rescula) in patients with open-angle glaucoma and suspected glaucoma. *Am. J. Ophthalmol.,* Oct; *134(4)*, 552-9.

[108] Nicolela, M. T., Buckley, A. R., Walman, B. E., Drance, S. M. (1996). A comparative study of the effects of timolol and latanoprost on blood flow velocity of the retrobulbar vessels. *Am. J. Ophthalmol.,* Dec; *122(6)*, 784-9.

[109] Inan, U. U., Ermis, S. S., Yucel, A. & Ozturk, F. (2003). The effects of latanoprost and brimonidine on blood flow velocity of the retrobulbar vessels: a 3-month clinical trial. *Acta Ophthalmol. Scand.,* Apr; *81(2)*, 155-60

[110] Harris, A., Migliardi, R., Rechtman, E., Cole, C. N., Yee, A. B. & Garzozi, H. J. (2003). Comparative analysis of the effects of dorzolamide and latanoprost on ocular hemodynamics in normal tension glaucoma patients. *Eur. J. Ophthalmol.,* Jan-Feb; *13(1)*, 24-31.

[111] McKibbin, M. & Menage, M. J. (1999). The effect of once-daily latanoprost on intraocular pressure and pulsatile ocular blood flow in normal tension glaucoma. *Eye., 13*, (Pt 1), 31-4.

[112] Vetrugno, M., Cantatore, F., Gigante, G. & Cardia, L. (1998). Latanoprost 0.005% in POAG: effects on IOP and ocular blood flow. *Acta Ophthalmol. Scand. Suppl., 227*, 40-1.

[113] Short-term (2004). effect of latanoprost on ocular circulation in ocular hypertension. Akarsu C, Bilgili YK, Taner P, Unal B, Ergin A. *Clin. Experiment. Ophthalmol.,* Aug; *32(4)*, 373-7.

[114] Whitson, J. T. (2002). Travoprost--a new prostaglandin analogue for the treatment of glaucoma. *Expert. Opin. Pharmacother.,* Jul, *3*, 965-77.

[115] Allemann, R., Flammer, J. & Haefliger, I. O. (2003). Absence of vasoactive properties of travoprost in isolated porcine ciliary arteries. *Klin. Monatsbl. Augenheilkd.,* Mar; *220(3)*, 152-5.

[116] Sakurai, M., Araie, M., Oshika, T., Mori, M., Masuda, K., Ueno, R. & Takase, M. (1991). Effects of topical application of UF-021, a novel prostaglandin derivative, on

aqueous humor dynamics in normal human eyes. *Jpn. J. Ophthalmol., 35*, 156-165.

[117] Makimoto, Y., Sugiyama, T., Kojima, S., Azuma, I. (2002). Long-term effect of topically applied isopropyl unoprostone on microcirculation in the human ocular fundus. *Jpn. J. Ophthalmol.*, Jan-Feb; *46(1)*, 31-5.

[118] Kojima, S., Sugiyama, T., Azuma, I., Konishi, N. & Fujii, H. (1997). Effect of topically applied isopropyl unoprostone on microcirculation in the human ocular fundus evaluated with a laser speckle microcirculation analyser. *Nippon Ganka Gakkai Zasshi.*, Jul; *101(7)*, 605-10.

[119] Devane, W. A., Hanus, L. & Breuer, A. (1992). Et al: Isolation and structure of a brain constituent that binds to the cannabinoid receptor. *Science, 258*, 1946-9.

[120] Cravatt, B. F., Prospero-Grcia, O., Siuzdak, G., et al: (1995). Chemical characterization of a family of brain lipids that induce sleep. *Science, 268*, 1506-9.

[121] Calignano, A., La Rana, G., Giuffrida, A. & Piomelli, D. (1998). Control of pain iniated by enodogenous cannabinoids. *Nature, 394*, 277-810.

[122] Woodward, D. F., Tang-Liu, D. D. S., Madhu, C., et al: (2000). Prostaglndin F2α (PGF2α) 1-ethanolamide: a pharmacologically unique local hormone biosynthesized from anandamide. 11th International Conference on *Advances in Prostaglandin and Leukotriene Research*, p27.

[123] Yu, M., Ives, D. & Ramesha, C. S. (1997). Syntheis of prostaglandin E2 ethanolamide from anatamide by cyclooxygenase-2. *J. Biol. Chem., 272*, 21181-6.

[124] Maihöfner, C., Schloetzer-Schrehardt, U., Gühring, H., et al: (1999). Expression of cyclooxygenase-2 in normal and glaucomatous eye [abstract*]. Invest. Ophthalmol. Vis. Sci., 40*, S827.

[125] Michels-Rautenstrauss, K., Mardin, C. Y., Schweitzer, D., et al: (2000). GL-CIA-locus: mutation screening in MYOC/TIGR and COX-2 of german POAG patients [abstract]. *Proc. Am. Soc. Human Genetics., 65*, 2731.

[126] Brubaker, R. F. (2001). Mechanism of action of bimatoprost (Lumigan). *Surv. Ophthalmol.*, May 45 Suppl 4, S347-51.

[127] Allemann, R., Flammer, J., Haefliger, I. O. (2003). Vasoactive properties of bimatoprost in isolated porcine ciliary arteries. *Klin. Monatsbl. Augenheilkd.*, Mar; *220(3)*, 161-4.

[128] Joyner, W. L., Gilmore, J. P. (1986). Tissue grafted into cheek pound of the hamser, in Baker CH, Nastuk WL (eds): *Microcirculatory Technology. Physical Techniques in Biology and Medicine*. Academic Press, London, 3-17

[129] Vetrugno, M., Cardascia, N., Arnese, L., Cantatore, F. & Sborgia, C. (2004). The effects on IOP and pulsatile ocular blood flow in patients with primary open angle glaucoma switched from timolol. 5th International Symposium on *ocular therapeutics (ISOPT)*, Montecarlo, Monaco, March, 11-14.

[130] Vetrugno, M., Cardascia, N., Arnese, L., Cantatore, F., Sborgia, C. (2004). The effects on IOP and pulsatile ocular blood flow of bimatoprost in patients with primary open angle glaucoma switched from timolol. 5th International Symposium on *ocular pharmacology and therapeutics*. March, Montecarlo.

[131] Ocular hemodynamics in normal tension glaucoma: effect of bimatoprost. Zeitz O, Matthiessen ET, Wiermann A, Reuss J, Richard G, Klemm M. *Klin. Monatsbl. Augenheilkd.* 2004, Jul; 221(7), 550-4.

[132] Coleman, R. A., Smith, W. L. & Narumiya, S. (1994). VII International Union of Pharmacology classification of prostanoids receptors: Proprieties, distribution, and structure of the receptors and their subtypes. *Pharmacol. Rev., 46*, 205-29.

INDEX

A

B

C

D

E

F

G

H

I

K

L

M

N

R

S

T

U

V

W

Y